A Shearwater Book

Nature's Restoration

ℭ

Nature's Restoration

PEOPLE AND PLACES ON THE FRONT LINES OF CONSERVATION

Peter Friederici

ॐ

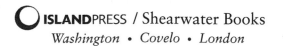
ISLANDPRESS / Shearwater Books
Washington • Covelo • London

A Shearwater Book
Published by Island Press

Copyright © 2006 Peter Friederici

SHEARWATER BOOKS is a trademark of
The Center for Resource Economics.

Library of Congress Cataloging-in-Publication data.
Friederici, Peter, 1963-
  Nature's restoration : people and places on the front lines of conservation / Peter Friederici.
      p. cm.
  Includes bibliographical references and index.
  ISBN 1-55963-085-x (cloth : alk. paper)
  1. Restoration ecology--United States. I. Title.
  QH104.F732 2006
  333.72—dc22

                                                  2005034289

British Cataloguing-in-Publication data available.

Printed on recycled, acid-free paper ✪

Design by Joyce C. Weston

Manufactured in the United States of America

10 9 8 7 6 5 4 3 2 1

For Michele, who was there

Contents

᠕

Nature's Restoration

One Man, Fifteen Acres, Forty Years

*O*ne sunny April afternoon in 2002 I accompanied David Wingate as he piloted a Boston Whaler, borrowed because his own was under repair, from the wharf on Nonsuch Island out to one of the only four islets in the world where cahows nest. The small motor that raises and lowers the outboard on his own boat had failed a day earlier, leaving Wingate stranded in the narrow channel between Nonsuch Island and Cooper's Point on Bermuda's mainland. He had ended up wading to Nonsuch—something he'd never thought possible, but fortunately the tide was extremely low, the water never more than chin deep.

His hair and curly beard had turned white, but at sixty-seven Wingate was fit and broad-chested. In his faded shorts and polo shirt and well-worn Top-Siders he looked like one of those gentlemen of leisure you find in ports around the world who spend much of their time messing about on boats. The impression was accurate enough: Wingate had retired two years earlier. The trips he took were still workaday matters, though—commutes, really, covering the short distance from Bermuda's mainland out to the islets where cahows nest or to Nonsuch, the nature preserve he'd begun working on forty years earlier. His piloting showed the effect of long familiarity with these waters: he pounded ahead ferociously, bouncing on the choppy swells. The cerulean of Bermuda's shallow waters was deep and saturated, as if the clear sky had fallen into it. We hung on as we rounded the south tip of Nonsuch, a ragged limestone point cloaked

1

with scrubby buttonwood trees, and bounded another few hundred yards to a small islet. No more than thirty feet high, the island was covered with short green grasses and succulent plants. Eroded by wind and waves, its limestone was as jagged as fresh lava. We circled the islet and anchored in its lee, only a few feet from shore.

Wingate was still limping just a bit from a recent knee surgery, but nonetheless he leaped up onto the islet's limestone rim. I followed him. He had been here thousands of times, and the weather was mild compared with that of many other days. Once, he recalled, the waves threw the boat so high that the outboard caught in the rocks; often he'd been soaked in cold rain; other times he'd been stranded, unable to return to Nonsuch or the mainland at all until the waters calmed. On many days he'd had to swim from the boat to shore, but today was easy and dry.

჻

DAVID WINGATE WOULD LIKELY not have been making this trip had he not taken a similar trip more than half a century earlier. It was 1951, and he was sixteen years old. That year Louis Mowbray, a Bermudian who was the curator of the island's aquarium, and Robert Cushman Murphy, an ornithologist from the American Museum of Natural History, teamed up to search for nests of the cahow, a striking black-and-white petrel, a seabird about the size of a crow, that was endemic to Bermuda. The cahow spent most of its time feeding and roosting on the surface of the Gulf Stream, hundreds of miles to the west; in winter it arrived on Bermuda and nested in deep holes in the ground. Like many other seabirds, it was entirely nocturnal near land, a tough bird to spot, although early settlers did see it, heard its clear whistled calls, and gave the species its onomatopoeic name.

It is easy to understand the appeal petrels have for birders and sailors. Of all the world's birds, none comes closer to slipping the bonds of land. Were it not for their need to find a solid place on which to lay an egg and raise a nestling, one suspects that petrels would never set foot on land at all, perhaps never even fly within sight of it. Petrels feed far out at sea, alighting on the water or dipping their bills into it while aloft to catch small squid and other aquatic animals. Because many such animals live deep in the ocean during the day and only rise near the surface at night, petrels have excellent night vision and feed most often by dark or at twilight. Specialized tubes that lie atop their bills and shield their nostrils may aid them in smelling prey for great distances. They can drink seawater and excrete its salt from glands above the bill, obviating any need for fresh water.

Above all, petrels are equipped to fly at sea. Long, pointed wings maximize the lift they gain from wind and from the updrafts that are created when waves push air upward. Many seabirds soar by skimming the updrafts a wing tip away from wave crests, but a petrel can particularly thrill observers by darting high into the air until it almost stalls, then accelerating downward until it disappears behind a wave crest, where it takes advantage of the lowered wind speed to accelerate and zoom upward again. Like a feathered embodiment of the ocean's ceaseless movement, it rides an invisible roller-coaster above the swells, moving leagues without needing to flap its wings, as attuned to its own restless element as the most streamlined tuna or shark.

Like any wave, these birds do come ashore at some point, and it is this need that has carved petrels into some two dozen distinct species, each associated with very particular breeding grounds. Petrels are awkward on land; they waddle about a bit like ducks, although they can climb branches and dig burrows with their clawed feet. Because they are especially vulnerable on the ground, many breed on islands more or less free of predators, and it is this restriction to particular islands that has led to the creation of so many species. The cahow evolved on Bermuda as the Jamaican petrel evolved on that island.

It was not grace, though, that lay behind the appeal of the cahow back in 1951. The species was an object of fascination because most Bermudians believed it extinct. They knew the cahow only from distant historic lore, as a symbol of the very early days when their land was new to human experience. In Bermuda it represented extinction, and the human excess of early colonial days, as effectively as the passenger pigeon did in North America.

Bermuda lies far out in the ocean, 586 miles from the nearest point of land, Cape Hatteras, North Carolina. Its history began when a great underwater volcano erupted some 110 million years ago and

created a mountainous island that was eroded back to sea level by
the waves. Corals grew on the lava shoals. When the sea level
dropped, waves pounded the corals into sand that was eventually
compressed into a porous limestone. When it rose, seawater eroded
sharp cliffs and fed more corals. A long series of rises and falls in sea
level created more corals, sand, rock, hills, and caves. By modern
times, Bermuda consisted of an archipelago of about 150 islands
over about twenty square miles—about the size of Manhattan—set
in a complex matrix of offshore reefs and shoals.

Life arrived. Bermuda was an experiment in what was able to
cross hundreds of miles of open ocean. What survived became, as
Wingate put it to me that April day, an "impoverished and eccentric
fauna." Whales spouted offshore, and bats passed by on migratory
routes, but not a single land mammal colonized the islands. Until
about 1505, that held true for humans too. As far as archaeologists
have been able to tell, the place was entirely uninhabited before
Europeans got there. Sea turtles abounded about the reefs, but the
only terrestrial reptile that made it was a skink whose eggs probably
arrived, by chance, in an upturned root wad or mass of vegetation
ripped out to sea by a riverine flood.

Birds arrived too. Draw a straight line between Nova Scotia and
Brazil, and it passes directly over Bermuda. Many land birds and
shorebirds follow migratory routes that take them over or near
Bermuda; many others are blown far out to sea during migration by
prevailing westerlies, especially in autumn, and make landfall on the
islands. Of all these visitors, some land birds remained to nest and
some evolved into unique island species: among others, there were
flightless rails, a large-billed finch, a crow, a stocky heron, a small
owl, a woodpecker, and a subspecies of the white-eyed vireo, a small
gray songbird with a catchy, lilting song.

Seabirds, of course, had no trouble getting to Bermuda. And
except for a few hawks and owls, there were no predators on the
islands, so they constituted a safe nesting place. Terns of at least three
species nested on the bare ground, as did short-tailed albatrosses.
Shearwaters and cahows settled in deep soil burrows, while white-
tailed tropicbirds chose rocky niches and crevices in coastal cliffs.
Surrounded by thousands of square miles of rich feeding waters,

these birds grew common. There were, it has been estimated, a half million pairs of cahows alone, nesting in burrows all over Bermuda, when a Spanish ship commanded by Juan de Bermúdez arrived offshore in the early years of the sixteenth century, carrying what were probably the first human eyes ever to behold the islands. The mild night air was filled with the eerie cries of birds and the churning of surf on reefs. Bermuda lay on the path galleons took home from the West Indies, but it was a treacherous place for sailing ships; the Spaniards called it "the Isle of Devils." They didn't settle, although one crew spent enough time ashore to drop off some hogs in hopes of providing provenance for future visitors, and another collected more than a thousand cahows on the island and salted and dried them for traveling fare.

Bermuda was, then, pretty close to a virgin land when the first settlers got there. Like many of the island's animals, they arrived by chance. The year was 1609; the ship was the *Sea Venture*, bound for Virginia from England; the commander was Sir George Somers, out to deliver settlers to that new colony. The *Sea Venture* carried 150 souls, all of whom miraculously managed to survive when the ship ran aground on offshore rocks.

When they made their way ashore, they found that Bermuda was more paradise than hell. "The fairies of the rocks were but flocks of birds," one account put it, "and all the devils that haunted the woods were but herds of swine." Sea turtles abounded in the shallows and laid masses of eggs on the sandy beaches. Fish teemed on the reefs. There were fat hogs to eat. Terns laid what one early chronicler, John Smith, called "infinite store of egges" on small islets. Almost a bother to the human intruders, cahows were wont to "light upon their shoulders as they went, and leggs as they satt, suffering themselves to be caught faster than they could be killed"; the birds "with their multitudes and tamenesse wearied the catcher with being caught." The dense woodland that covered the islands was low and scrubby: not majestic, but useful. A sort of olive dangled from lime green branches; sweet reddish fruits reminiscent of pomegranates clustered on the prickly pears. The palmetto berries could be fermented into a passable wine. Cedar trees provided splendid timber, and the fibrous palmetto leaves could be used for thatch and rope.

The land was rich, the climate mild, the promise of the place great, and there were no inconvenient native people in the way; to the inadvertent visitors, it seemed as though the hand of Providence had placed this great storehouse in the middle of the ocean's wastes specifically for them. When after nine months the expedition left in two ships built of lumber salvaged from the *Sea Venture*, two men opted to stay behind—the island's first settlers.

Somers returned for them later in 1610, but died on the islands, where his heart lies buried today; the rest of his body was taken back to England. One of his men elected to remain on Bermuda with the two early castaways. For two years the trio led a prelapsarian life, growing corn, fishing, eating pork and birds' eggs, brewing palmetto wine. When a 150-pound chunk of ambergris washed up on a beach they thought they were rich, but soon quarreled over it and nearly fought a duel. They had decided to abandon their Edenic life and attempt to return home by building a boat and sailing to Newfoundland when another expedition arrived from England in 1612 with a load of settlers who, this time, were backed by wealthy investors and planned to stay on Bermuda.

It was the end of an idyll. The Bermudians soon found their mother lode of ambergris whittled away by savvier businessmen back home. Within a few years rats arrived from England in a load of meal and multiplied so rapidly that they threatened to eat all the newly planted grains. Cats, hastily shipped in, didn't help. A series of governors ordered the island burned to eradicate the rats. That didn't work either, but it did destroy great swaths of the native woodland.

The settlers quickly revealed that if Bermuda was not paradise it was due largely to their own actions. They slaughtered sea turtles; collected turtle and bird eggs by the thousands; ate the fat cahows well past the point of gluttony, as Smith related: "How monstrous was it to see, how greedily euery thing was swallowed down; how incredible to speake, how many dozen of thoes poore silly creatures, that euen offered themselves to the slaughter, wer tumbled downe into their bottomelesse mawes." The rats, cats, and hogs had their own appetites for animals and for eggs; hogs and goats rooted out vegetation. Cedars and yellowwoods were cut for timber. Tobacco was planted as a cash crop, and slaves were imported to work it.

For the native species that had evolved on Bermuda, without predators, it was devastation. The native crow and owl and heron and finch and probably other birds disappeared quickly; sea turtles dwindled with such alacrity that an act to protect the few survivors was passed only eight years after the colonizing expedition arrived. It was the first conservation legislation in the New World. Seabirds were protected too, but this proclamation, Smith wrote, was "ouerlate," for the birds "wer almost all of them killed and scared awaye very improuidently by fire, diggeinge, stoneinge, and all kinds of murtheringes." The cahow, as far as Bermudians could tell, had vanished. Its nest burrows were empty, the night air still. It was gone so quickly and so thoroughly that no specimens or even accurate illustrations existed. Centuries later, every Bermudian knew that cahows had been common and then vanished, but no one knew exactly what they had looked like. Ornithologists speculated about the genus to which the cahow had belonged. Was it a petrel? A shearwater? No one knew for certain.

The Bermudians quickly found out what settlers throughout North America would continue to learn, time and again, in a pattern that was nothing if not predictable but seemed to its pupils always a surprise: what appeared to be endless natural resources "were really only skin deep," as one nineteenth-century writer noted. The Bermudians, of course, survived this loss, but they did it through wholesale neglect of the demolished natural resources that had fueled the gluttony of those early years. They grew tobacco and other nonnative crops, they fished, they traded, and in some cases they grew wealthy through such marginally legal practices as privateering.

One resource, though, truly did seem inexhaustible. Once the tale of the *Sea Venture*'s wreck reached England, everybody knew about it from spoken stories or written accounts: the event was the *Perfect Storm* of its day. One or more of those accounts provided the source material for Shakespeare's *The Tempest*—and Bermuda became a byword for an impossibly remote place, bathed by mild winds, benevolently enchanted and removed from the woes of the modern world. It was paradise—and a paradise, what's more, to which the English felt themselves specifically invited by Providence. John Smith pointed out that its latitude, climatically and perhaps spiritually too, was almost precisely that of "Jerusalem, which is a clime of ye sweetest and most pleaseinge temper of all others." In the 1650s, Andrew Marvell encapsulated the common perception in a famous poem called "Bermudas," about a place enameled with "eternal spring" where food more or less drops from the air and the trees: fowls, figs, melons, and apples, all ripe and ready for those who would sing the praises of the hand that placed them there. In gray England or cold Boston, the vision was far more potent than any reality. It didn't matter to readers that much of the islands' abundance was already gone by then; the benevolent climate, and reputation, remained.

Those who oversee the industry that in the twentieth century came to dominate the islands' economy and ecology—tourism—don't mind the paradisical reputation at all. Nor do most of those who visit, who can sunbathe under coconut palms, watch the warm breeze toss the graceful tops of casuarina or Norfolk Island pines, admire the brightly blooming bougainvilleas and oleanders, play golf on manicured lawns of short-cropped Bermuda grass and Kentucky bluegrass, smell the fragrant Easter lilies whose export to the East Coast of the United States was once a major industry, see green and brown anoles scurry among the noisy dead fronds of Chinese fan palms, listen to the shrieks of kiskadee flycatchers during the day and the clear loud tones of whistling frogs during the night, and in short immerse themselves in all the sensations of a splendid subtropical getaway salved by the Gulf Stream, rimmed by coral beaches, and populated by friendly people with endearing accents. The only trouble is, not a single one of those species—plant, animal, or human—gives any idea of what Bermuda was like when Somers' men found their earthly paradise, for the simple reason that they were all brought to the islands from elsewhere. Even Bermuda grass is not native to Bermuda. And David Wingate can make the argument that many of these imported species make it more difficult for the relatively few native plants and animals, most of which are more or less invisible to the tourists and even to native Bermudians, to survive. So he has, for more than forty years now, been a tireless advocate for the Bermuda palmetto and the olivewood, the white-eyed vireo and the Bermuda skink. And, above all, for the cahow.

⅌

By 1951, Louis Mowbray was convinced that the cahow was not quite extinct. His father, a naturalist and the founder of the aquarium, had in 1906 collected a living specimen of an otherwise unknown petrel from an islet near Castle Harbour at Bermuda's east end; a fledgling flew into a lighthouse window on the eastern tip of Bermuda in 1935; in 1945 a dead individual washed up on a beach at Cooper's Point, which forms the east edge of Castle Harbour; and fishermen still reported hearing odd calls at night in this area. By then Bermuda's quiet had been shattered. Cars had arrived en masse after World War II. More dramatically, the U.S. Navy had built an airstrip at Cooper's Point at the war's onset, and it remained a busy place throughout the Cold War, with bright lights piercing the night and long-range nuclear bombers and submarine-tracking aircraft arriving and departing all night long. Once the space program began, NASA also came to operate a brightly lit facility there that tracked the flight of spacecraft taking off from Florida.

In the late winter and early spring of 1951 Mowbray and Robert Cushman Murphy poked around the Castle Harbour islands, peering into deep crevices eroded into the limestone and looking for signs of occupancy. Finally, one day Mowbray attached a wire noose to the end of a long pole, fished into a likely looking burrow, and pulled out a striking-looking black-and-white bird with a hooked black beak. "By Gad, the cahow!" he exclaimed.

David Wingate, sixteen years old, had accompanied the scientists that day because they knew of his interest in birds. Born to Scottish immigrant parents, he was a native Bermudian. Alone among his peers, he was "a spontaneous naturalist," in his words, who had already taught himself to identify each of the dozens of species of North American warblers that regularly winter on Bermuda, along with the islands' many other birds. He too had become fascinated by the lore of the cahow. As the men examined their find he realized what he wanted to do with his life. That day launched him on a course to Cornell University, where he studied zoology. By 1958 he was back in Bermuda, a freshly minted college graduate trained for a job that didn't exist there: conservationist. Under Mowbray's guidance, and with a grant from the New York Zoological Society, he was able to begin to work on the cahow's conservation. Within a decade he'd become the islands' reigning expert on flora and fauna and had talked himself into a newly created job: conservation officer for Bermuda, in charge of preserving endemic and rare species such as the cahow and the Bermuda skink. He was the only qualified candidate.

His first task was to preserve the few remaining cahows, which, as had quickly become evident in 1951, were critically endangered. Mowbray and Murphy, with Wingate's aid, eventually found eight nests that year. They suspected that predation of eggs by rats was the greatest threat to the species. On tiny islands that was an easy problem to solve with the application of a bit of rat poison. Within a few days of the first discovery the eggs of the cahows, one per pair, had hatched. Then an unexpected problem arose. Within a few more days the white-tailed tropicbirds had returned to Bermuda for their nesting season. Tropicbirds nest in rocky cavities just like those occupied by the cahows. Longtails, as Bermudians call them, are striking white seabirds with black highlights and long white tail feathers that stream gracefully in the marine wind. They fly along the coastline in pairs and trios, incessantly calling *kip-kip-kip*. These lovely, lively birds live throughout tropical waters and nowhere nest farther north than on Bermuda, so it is perhaps appropriate that they came to serve as the islands' unofficial emblem, a vivid symbol of the Gulf Stream that stokes the tourist economy.

The researchers who rediscovered the cahow, though, were not sanguine about them. Within days of the rediscovery, pairs of tropicbirds had entered every cahow nest but one, killed each chick, and taken over. When the adult cahows returned, at night, they found themselves expelled from their own nests. Clearly, something would have to be done, and soon, but what?

Wingate knew that a potential solution lay at hand. A young American biologist, Richard Thorsell, had worked with the cahows a bit in the mid-1950s under the tutelage of the well-connected conservationist Richard Pough. They had realized that cahows are a bit smaller than tropicbirds. Pough had suggested building a cavity entrance that would allow cahows but not tropicbirds to enter, and Thorsell had experimented with such a design. Through trial and error, Wingate refined the design and figured out the precise oval size

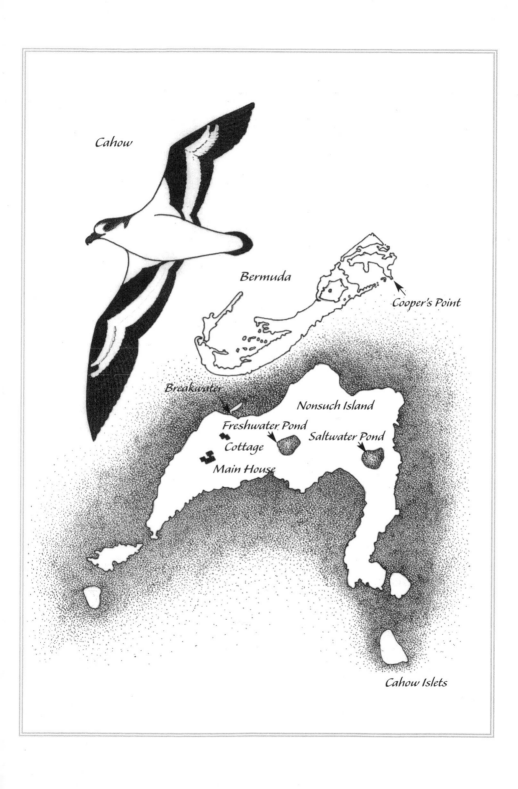

Cahow

Bermuda

Cooper's Point

Breakwater

Nonsuch Island

Freshwater Pond

Saltwater Pond

Cottage

Main House

Cahow Islets

and shape of a cavity that would allow cahows but not tropicbirds to enter. By cutting such an oval in small sheets of plywood and placing the plywood in front of active nest burrows, he hoped to prevent tropicbirds from bullying their way in. But the tropicbirds, like the cahows, had strong site fidelity: each spring they returned to exactly the same nest site and pushed their way in. Wingate had to wait on the islands until they returned, catch them by hand—helped along by their helplessness on land—and dispatch them. Once those pairs were gone, new tropicbirds didn't bother the cahow nests outfitted with what Wingate came to call "bafflers." Cahows nesting at those sites began successfully fledging young.

Still, Wingate could not understand how the species had managed to survive all those centuries. Of the eight nests known, seven had been subject to tropicbird invasion, and they contained piles of bones of nestling cahows; clearly, that had been going on for a long time. How had the species persisted so long? By 1960 it had become apparent that adult cahows live a long time, for the same birds returned year after year. They did not have to reproduce successfully each year to ensure the species' survival. Still, only one successful nest could not explain the persistence of the species. After the initial discovery Wingate and the other ornithologists had spent the night on the first nesting islet and had observed an estimated thirty cahows flitting in several hours after sunset, uttering soft call notes. Some were probably subadults not yet ready to breed; like other petrels, cahows do not nest until they are about five or six years old. Nonetheless, he suspected that more nests existed somewhere. He began spending more time on the islets at night.

It was tricky work. Cahows lay their eggs in midwinter. Landing could be treacherous. The islets were wind-whipped and offered no shelter. And it took a long time before Wingate spotted a single cahow. The birds seemed to avoid returning to their nests entirely on clear and calm nights. Night after night he kept watch, seeing nothing.

"I began to think I was going crazy," he said. "All the daytime evidence suggested something would happen, but I'd go out there at night and not see a thing." One night a sudden gale arose, and he was stranded by the wind and waves. Then they came: petrels flying in, graceful on the strong wind, and landing and disappearing into cavities. He'd found them. In the winters of 1960 and 1961 he found eleven new nest sites. And he learned why cahows could nest there: thanks to quirks of topography, the burrows naturally prevented competition with tropicbirds.

With bafflers built at the nest cavities that needed them and with new nests located, Wingate watched as the cahows experienced increased nesting success in the 1960s. Yet only four islets were large enough to provide dry burrows and small enough to lack rats. No more than thirty-five appropriate natural burrows existed on them. In total, the islets encompassed only about two acres. Cahows, it was clear, remained on the brink. Without new nesting sites, Wingate wondered, how could the population grow?

ॐ

THE TWO-PART ANSWER TO THAT QUESTION has occupied much of David Wingate's life and was very much on my mind as we walked up the islet, striding cautiously on the sharp limestone. Just before us, we could see an array of what looked like cement mole tunnels built just above the lowest limit of the vegetation. It was the first answer to the question of the cahow's survival, one that will be familiar to anyone who has ever seen a purple martin colony or a bluebird box: artificial nests.

After a few years of observing the natural nests Wingate had a good feel for what cahows look for in a nest site: the ideal place is a cavity reached by a passage long enough to be truly dark, with a bit of open space before its entrance, hard by the nests of neighbors. It was not difficult to design an artificial version, although it was hard work ferrying the building materials out to the islets: wire mesh and 115 buckets of concrete, each one hoisted up from a heaving boat. Then there was the labor of excavating the six- to eight-foot trench each burrow required out of hard rock, and the work of molding the mesh into a framework and pouring the concrete over it.

Each artificial nest begins with a gray-painted plywood baffler that leads into a long passage that curves its way back to a round nest chamber a bit larger than a basketball. Wingate tried to leave a bare patch of dirt in front of each baffler—the cahows seem to prefer it, and it allows him, sometimes, to spot signs of activity. As we walked among the concrete nests, stepping carefully among sharp limestone, rounded concrete passages, and trailing, succulent halophytes that grew no more than a few inches high, he stopped in front of each to scrutinize the dirt. In some cases he had placed tiny twigs upright in front of the entrances so that like a detective he could deduce a bird's passage. "It's so difficult to read the evidence," he muttered. Still, in front of several entrances he saw a scuff in the dirt, a bit of whitewash, or a toppled twig that indicated that seabirds had been there.

Each nest chamber is topped with a round lid that can be removed for an easy look into the nest. Wingate walked to each, removed the heavy rock or two that he'd laid on each lid to keep it from being blown off by strong winds, lifted the lid, and looked inside. Once, twice, three times he murmured, "There's one," and he looked more closely.

I am an avid birder and had come a long way to see this place and this bird, but this method seemed a bit like cheating: it was too easy a way to see a bird so stealthy, so bound to an enormous and—to us—trackless habitat, so good at making itself scarce. The cahow chicks were formless round masses of dark gray downy feathers, as large as softballs and as unmoving. Cahows only return to feed their nestlings every few nights, and when they do they manage to regurgitate a great quantity of squid and other sea creatures. The nestlings gorge themselves and then sit and wait, in the darkness, for another night and another feeding spree. They grow fat and are nothing but raw potential.

"A long way from the superb flying machines they become, aren't they?" Wingate commented as he carefully fitted the first lid back on. He jotted the results in a tiny, tattered notebook: so many nests examined, so many nests occupied. One nest contained a tropicbird egg, as large as a chicken's. Somehow the tropicbirds had bulled their way in, but since there were no cahows using that cavity Wingate left it there. Tropicbirds, too, have grown scarcer in Bermuda. The construction of houses atop sea cliffs has destroyed some of their nesting habitat, as has a series of severe storms that have eroded the shoreline, a trend Wingate ascribes to global warming and a resulting rise in sea level. In the 1990s he went so far as to begin the construction of artificial tropicbird nest burrows that can be seen rising like small concrete igloos atop some of Bermuda's sea cliffs.

By 2002 Wingate had built more than sixty artificial burrows and knew of sixty-five nesting pairs of cahows. He estimated that the total population had risen above two hundred individuals—a slow rise, but a steady one too. They had survived against stark odds—not only the long centuries of presumed extinction and utter neglect, but also the global contamination by DDT in the 1960s, which caused thinning of eggshells and had dramatically lowered reproductive success in the cahow as in many other predatory birds; the depredations of a vagrant snowy owl in early 1987, which killed five cahows before Wingate reluctantly shot it on Nonsuch; and the lights of the airbase and NASA station on nearby Cooper's Point, which threatened to confuse the night-flying adults. The cahow, it appeared, was safe, at least as safe as it had been anytime in the last 380 years, thanks to an immense amount of work on Wingate's part. In addition to the initial surveys and the construction work, there was the regular monitoring; he insisted on motoring out to the nest islets every two to four days during the long nesting season, from January until May or June, to check that the bafflers were working properly. He did not want to find a cahow chick or adult stuck in a burrow, trapped by a baffler that happened to be a bit too small. So year after year this became his ritual: boating out to the islets, in fair weather and foul, leaping ashore, and checking the nests. He was a godparent to every fledgling cahow, and found nothing more fulfilling than being out on one of the islets some late spring evening when a young cahow emerged from its carefully watched nest burrow and launched itself a bit unsteadily over the dark sea, trusting for the first time in its buoyancy and its lithe wings. More than once Wingate found himself rescuing a bird that, not quite ready, plunged to the ground or into the swells instead. But all this tireless effort played out day after day and night after night over so many years, all these thousands of boat trips and long days and late nights—all this, as it turned out, was only the smaller part of his labors.

☙

As HE WORKED ON THE ISLETS DURING THOSE EARLY YEARS, Nonsuch Island loomed in Wingate's thoughts. He motored past its craggy shore on every trip. It was scarcely larger than a typical city block, and it was desolate. Because of its isolation, it had served as a yellow-fever quarantine site in the late nineteenth century. In the 1930s the naturalist William Beebe had used it as a research head-quarters as he studied the reef and deep-sea fishes around Bermuda. Later it became a reform school for boys, but its exposure to southerly storms and its lack of a safe anchorage ended that too. In the 1950s all its native cedar trees, like the great majority of the cedars on Bermuda, had died of a blight inadvertently imported from California. The goats kept by a caretaker had eaten much of the other native vegetation, including the shrubs that such birds as white-eyed vireos and gray catbirds required for nesting; much of the soil had washed into the ocean. It was, Wingate said, "a desert island," where the sturdy snags of the dead cedars slowly bleached moon white and stood starkly over low grass and scrubby lantana, ghostly reminders of a richer past.

It was because of the cahows that Wingate came to see the same potential in the dead trees, and the wasted island, as he saw in the portly chicks sitting so stolidly in their nest burrows. The cahow islets, he knew, would never support large numbers of nesting seabirds; they were too small and too exposed. They also never could provide soil deep enough for nest burrows. Above all, Wingate wanted to see cahows not only surviving, but nesting as they had in presettlement times, in burrows they'd excavated out of deep soil, in places tropicbirds would not go. That could only happen on a larger, well-vegetated island. And Nonsuch was perfect for this purpose because it was only a few hundred yards from existing cahow nests, yet small enough and far enough from the mainland that it might be possible to keep rats, cats, and other problematic animals in check. He estimated that a thousand pairs could excavate their own bur-rows on Nonsuch and that the cahow population could grow far larger still with the construction of artificial burrows there. It was the ideal and only long-term solution to the species' nesting problems.

It is breathtaking at the remove of forty years to think of a man setting out to revegetate an entire island just to create and shield soil so that birds that have almost vanished can proliferate and dig holes in it, but Wingate was young and energetic and determined and enraptured with his vision of the cahow's future, and perhaps a little ignorant of how much work he was going to have to do. "I didn't really worry about it," he said, "because it was so pleasurable."

He soon came to dream even more broadly. During his initial surveys he noticed that the Bermuda skink, the half-foot-long lizard that was the islands' only native land reptile, was common on Nonsuch. It had become rare elsewhere on Bermuda as land was developed and as such nonnative species as anole lizards and kiskadees had been introduced. On Nonsuch, the skink was forty times more abundant than on the mainland. "It dawned on me," he said, "that Nonsuch was a sanctuary for more than the cahow. At that point I said, 'Why not make this a sanctuary for *all* the native heritage of Bermuda and call it a living museum?'"

Nonsuch was a forsaken place, and no one else much cared what he did there. He talked the government into declaring it, along with the cahow islets, a wildlife sanctuary. He gained permission to move out to the island and live as a caretaker in the old yellow-fever hospital, and he moved out to the half-derelict building with his wife and daughter in 1962. He began weeding out the nonnative species, beginning with the rats and then moving on to the vegetation. As with the rediscovery of the cahow, he arrived in the nick of time; most of the nonnative plants that had come to dominate Bermuda's vegetation had not yet reached Nonsuch. He weeded intensively during his first years on the island, and within a year had cleared enough space to begin planting native trees and shrubs.

That proved a big challenge. The native cedars were almost all gone, and other species, having been neglected for landscaping in Bermuda while nonnative species such as oleanders and Indian laurels and casuarinas were widely planted, were hard to come by. Wingate found what he could and transferred boatload after boatload of seedlings from the mainland to Nonsuch, then propagated others on the island. He planted seedlings by the thousands without much certainty about their specific habitat requirements. Although only fifteen acres in extent, Nonsuch has a variety of habitats: beaches, sea cliffs and rock outcrops, upland ridges, a small interior valley with relatively deep and moist soil. Which species would best grow where? Native species were little known; no one could answer that question. So Wingate planted palmettos, olivewoods, hackberries, dogwoods, bay grapes, and others willy-nilly, hoping that their success or failure would answer the question for him.

Before it could, though, he realized that the native plants would require windbreaks. Nonsuch is not a large island, and it is exposed to the full brunt of southerly storms, especially hurricanes, that drive salt spray all across it. So Wingate did what people all over Bermuda were doing in the wake of the cedars' demise: he planted fast-growing, salt-tolerant, but nonnative casuarinas, which originally came from Australia, and tamarisks, native to Eurasia. He imagined them as a ring around the island inside which the native species could grow.

The early years were unprepossessing. The tamarisks and casuarinas did grow quickly but were labor-intensive. Casuarinas in particular tend to grow tall and wispy and not make very good windbreaks, so Wingate found himself needing to top them every couple of years to make them bushier. The native plants grew terribly slowly. The wind swept across the little island. Maintaining the sprawling old house took as much time, if not more, as maintaining the island.

By the mid-1960s, too, the cahow population came to stagnate as reproduction declined due to DDT poisoning. The pesticide wasn't used in Bermuda, but the petrels were picking it up from their prey as they foraged over the Gulf Stream. It was dispiriting. Perhaps all the work on the little islets might be for naught, Wingate thought—but then in 1971 DDT use was banned in the United States. The reproductive rate increased. Nonsuch, though, still looked like a scrubby wasteland. Few appreciated his efforts. One visiting government official muttered, "What is he culling and replanting native flora for when he ought to be planting pretty things like the poinciana to make a nice garden for the tourists?"

From there things got worse. By 1972 a scale-resistant strain of Bermuda cedars had been found and propagated, and Wingate was able to plant six hundred seedlings on Nonsuch. Most of them died. The following year Wingate's wife, Anita, died in a fire in their Nonsuch house. In his grief he was left raising two daughters by himself—ferrying them to the mainland in time for school before beginning his workday, carrying in the groceries and everything else the family needed by boat, keeping house, helping with the homework, and trying to integrate all that with his trips to the cahow islets and his upkeep of the island and the cumbersome old house. (He later remarried and now has three daughters.)

Anita was buried in the old yellow-fever cemetery in the island's sheltered vale, where the soil was deepest and the old cedars had grown largest; some of the snags there were a foot and a half in diameter. In her honor, Wingate began planting yellowwood trees, a rare, slow-growing endemic species that was difficult to propagate but whose wood was once prized for lumber. They took. Out of the necessity of raising a family by himself, though, his work suffered. He neglected to top the casuarina trees, and some of them began producing enough shade to kill the native plants growing under them. As a result, he finally had to girdle some of the trees to kill them. A full third of the palmettos he'd planted turned out to be a species imported from Puerto Rico, rather than the native variety. Doggedly, he cut them down and started over.

The natives were still growing terribly slowly. Wingate added fertilizer, improving their growth but also attracting a plague of land crabs that undermined the soil and devoured the plantings. After ten years, it appeared that all his work might be for nothing. Having suffered personal tragedy and what looked like professional failure on Nonsuch, with little support for his work from his superiors, he might have been excused had he washed his hands of the place and devoted his energies entirely to the cahow islets. Instead he dug in.

❧

THE LAND CRABS WERE NATIVE, and Wingate was quite familiar with them. On Bermuda, as on many oceanic islands, they play more or less the same role that rodents do on the continents, cleaning up detritus and eating small plants. Gardeners and golf course managers loathed them because they dug deep burrows and ravaged plants. Now Wingate found them honeycombing Nonsuch Island and eating his hard-won seedlings. Instead of reaching for the pesticide can as the golf course managers did, though, he reached for his history books. He'd read enough about Bermuda's past to know that a crab-eating heron had once lived there. Records of the early settlers and even of later naturalists mentioned a stocky heron, very similar to the yellow-crowned night heron of the U.S. mainland, that had apparently drifted unnoticed into extinction on Bermuda. In fact, yellow-crowned night herons still regularly showed up on Bermuda in winter as vagrants from the mainland, but did not remain to nest. Could, he wondered, a breeding population of these birds be established on Nonsuch? It would be a much more environmentally friendly—and elegant—means of controlling the crabs, and a step toward restoring the island to what it was before Sir George Somers stumbled ashore.

To attract herons, Nonsuch would need a pond, or better yet, two, Wingate thought. The island had two swales where ponds could be built, one in its center that could become a freshwater pond and one just behind its largest beach that could be brackish. Ponds, too, would further the living museum idea by giving visitors a more comprehensive picture of what presettlement Bermuda had been like. He raised money and, in 1975, excavated the ponds. He transplanted to them such native plants as cattails and rushes as well as an endemic species of killifish. He was gratified to see that these plants thrived quickly, unlike those in the would-be woodlands. He was also glad to see the ponds become bird magnets that attracted wintering and migrant songbirds as well as waterfowl.

Now he was ready for the herons. From 1976 to 1978 Wingate gathered forty-four nestling night herons from rookeries in Tampa Bay. He raised them on Nonsuch, where he fed them on a diet of crabs. They survived and became independent. Soon the paths on Nonsuch were littered with a crunchy detritus of carapaces and pincers.

At first the herons did not remain. They flew off to Bermuda's mainland, where they began nesting in a tract of woods and sinkholes not far from Nonsuch. Within a few years the population swelled. At last some returned to Nonsuch to nest in the buttonwood trees on the far side of the saltwater pond.

Wingate was ecstatic. The appetites of the herons saved his plantings. As a result, he was eventually able to reintroduce other native plants to the island, such as the very rare Bermuda sedge. "The herons clicked right back into place," he mused. "They were instantly successful. Who could have predicted that growing a sedge depended on a heron?" The herons also ended up winning him friends on the rest of Bermuda as golf course managers noticed a substantial decline in crab numbers. They didn't give a whit about sedges, but were quite pleased to save on pesticides.

The success of the heron reintroduction was only one of Wingate's ecological interventions with the island's animal life. It was, in a way, simply a more thoughtful echo of what had been going on in Bermuda for many generations, because the history of the islands could be viewed as a large-scale and long-term experiment in introducing new species of plants and animals to a small piece of land. From the time the Spaniards dropped off their hogs, people had been actively engaged in altering Bermuda's flora and fauna for their own purposes, with results that were mixed but most often devastating to the native species.

There were, for example, the anoles, several species of insect-eating lizards that were introduced from the Caribbean to control Bermuda's flies. That worked pretty well. Bermudians, however, blessed with their benevolent climate, also brought in citrus trees, which in time came to be infested with citrus scale insects. Taking a cue from California, they imported ladybugs to devour the insect pests. That worked also, but then someone noticed that anoles love to feed on ladybugs. What could be done about them? Someone who'd spent time in the Caribbean recommended introducing kiskadees, which are large, aggressive, and noisy yellow-breasted flycatchers that like to eat lizards.

By now it was 1957 and David Wingate was in college learning about how introduced species can threaten natives, especially on small islands. He wrote back home, pleading with the island government not to import kiskadees, but to no avail. They arrived and were released. They did set about devouring anoles, although not enough to make much of a dent in those species' enormous populations. They also began eating the eggs and young of eastern bluebirds and white-eyed vireos, both uncommon species, as well as rare Bermuda skinks. They were a disaster, but by the time the authorities so realized, it was too late to do anything about them. It was, Wingate said, "the end of a long chain of stupidity." Today they are among the most common and conspicuous birds on Bermuda, and many tourists and even Bermudians are not aware that their shrieking alarm calls are a landscape feature even newer than the whining of the ubiquitous motorbikes.

Bermuda is an interesting place to try to define precisely what is natural and what is not, a matter of considerable concern to those involved in ecological restoration and one about as easy to resolve as the old debate about angels dancing on the head of a pin. Bermuda before the arrival of the first Spanish sailors was absolutely natural in the sense that human intention and effort had played no role in determining what lived there; as far as anyone has been able to tell, indigenous people never set foot there. But what then? Many species were explicitly brought in by people, either deliberately or accidentally: rats and cats, pigs and goats, house sparrows and anoles, oleanders and banana trees and coconut palms, rabbits and Easter lilies, Indian laurels and Norfolk Island pines and Chinese fan palms. These introductions were certainly "natural" according to the laws of human nature, but not "natural" in the same way as was the presence of cahows, white-tailed tropicbirds, and Bermuda cedars, all of which had made it to the island entirely on their own. Then the picture becomes more muddied still. Starlings did not nest in Bermuda until the 1950s, having arrived on their own from North America, but they were in North America only because they had been introduced there from Europe in the late nineteenth century. Mourning doves began nesting in Bermuda about the same time, also having arrived on their own, but then they were native to North America. So David Wingate considers the doves a more or less natural addition to the islands' fauna, whereas he considers starlings an artifact of human intent, and a problem.

Through the 1960s and 1970s Wingate had mixed success in trying to eliminate nonnatives and promote natives on Nonsuch. He wanted to get rid of the anoles, but his early attempts to catch them all by hand had failed. As the trees grew, the lizards proliferated, and he realized he could never get rid of them all. Birds moved in too. Along with native gray catbirds were nonnative cardinals and European goldfinches, both species that were introduced to Bermuda long ago for their beautiful plumage and song. But they did not appear to be competing with any native species, and he did not see any reason to get rid of them. His living museum would be exactly that—living and changing, not a static diorama that attempted to slavishly copy what Bermuda was like in 1500.

Wingate tried to reintroduce sea turtles, too, which still appear regularly in Bermuda's waters but have not nested on its beaches since the overexploitation of the early colonial days. From the late 1960s into the late 1970s he had eggs of green sea turtles flown in from Tortuguero, Costa Rica, site of a huge nesting area, and buried them on Nonsuch's South Beach. Some sixteen thousand hatched, and he hoped that they would imprint on the beach as their natal ground and return when it came time to breed. So far they have not returned, but he has since learned that it may take anywhere between thirty and fifty years before they return to nest. He is still hopeful.

He had more immediate success with the white-eyed vireo, a native songbird that had managed to survive on Bermuda's mainland because it nests in a wide variety of shrubs. It was a different sub-species than that found in North America—tamer, as is common of island species, and reluctant to fly over open water. It had disappeared on Nonsuch when a 1963 hurricane ripped up what was left of the island's shrubs. By 1972 Wingate thought the vegetation on Nonsuch was beginning to grow sufficiently shrubby and dense for them, but he surmised that they likely would not fly across even the few hundred yards that separated Bermuda from Cooper's Point. He captured three pairs on the nearby mainland and released them. They took hold and reproduced. Wingate was pleased that they quickly seemed to grow even tamer on Nonsuch than on Bermuda proper. Along with the success of the heron introduction it was a sign, he thought, that things were reverting to the way they should be on a tiny island.

⁂

THE QUESTION OF WHAT IS NATIVE and what is not was very much on my mind as we headed out to Nonsuch Island, largely because Wingate kept pointing out the myriad nonnative plants along the way. We stepped into Jeremy Madeiros' Whaler at a tiny wharf in Tuckertown, which has some of the most exclusive real estate in all Bermuda. Madeiros is a genial young Bermudian who became Wingate's successor as conservation officer when the older man reached Bermuda's mandatory retirement age. With us also was a crew of four men who were working with Madeiros at constructing a toad barrier around the freshwater pond on Nonsuch. It was another attempt to deal with an invasive species that might yet prove crucial not only to the island's overall health, but also very specifically to the conservation of the cahow.

Across the cove Wingate pointed out Michael Bloomberg's new palazzo, only one of a series of mansions set among palm trees and elaborately manicured lawns. The sprawling multimillion-dollar houses were new variations on the old Bermuda theme: white or pastel walls, roofs built of stepped and whitewashed limestone slabs, with a narrow, three-inch-high retaining barrier near their bases that funnels rainwater into a cistern; because Bermuda is built of porous limestone, it has no streams and its people must rely on water collected in this way.

Wingate fulminated equally about the people who build these extravagant vacation houses—"It's obscene for people to build thirty-million-dollar houses when there is so much hunger in the world, don't you think?"—and about the landscape they were set in—"No part of Bermuda is undisturbed. It's like a city without a country. Only one-third is at all rural. The largest open space is less than one hundred acres. And ninety-five percent of the vegetation is nonnative: Brazil pepper, Chinese fan palms, Indian laurel, casuarinas."

Jammed onto the small boat, we bounced across the riffled waters of Castle Harbour, past more opulent pastel vacation houses and cliffside pools and casuarinas. The airport sprawled before us on the sound's far side, a hotel to our left. And then, to our right, appeared some of the cahow islets, too small and low for trees, and then a larger island marked by a high limestone cliff. This was Nonsuch. Above the cliff swirled tropicbirds, flying in pairs and trios, calling *kip-kip-kip*.

We passed under the cliff, rounded a point greenly clad with vegetation, and docked at a cove in which William Beebe had had an old eighty-foot hulk sunk more than seventy years ago so as to provide safer anchorage; a bright blue parrotfish swam through its rusted ribs. We stepped onto the concrete wharf and then through a limestone passageway that led upward onto the island's relatively flat top.

Mourning doves called. We crunched over the saucer-sized fallen leaves of gangly bay grape trees, which cast an open and airy shade.

In the height of the bay grapes and of the other trees that we soon passed under—rustling palmettos and dense, yellow-green olive-woods—it was possible to read Wingate's success at reestablishing the native vegetation. Once the night herons began controlling the crab population, the native trees and shrubs really came into their own, and by 1980 they formed a dense woodland over much of the island. We walked on a mown path among the twenty-foot-high trees and among masses of shrubs and low forbs that I could not recognize. What Wingate described as native species, crammed tightly together, covered the forest floor: turnera, Bermuda sedge, turkey berry, the pretty poinsettia called Joseph's coat, the slender purple-flowered Bermudiana—Bermuda's national flower—and grasses. Wingate hadn't paid much attention to these species for decades, focusing instead on the trees and shrubs. In recent years, though, he'd had help from volunteers who had planted some of them; others had made it out to the island on their own.

What did interest Wingate, passionately, were the Brazil pepper seedlings. My family and I had taken a glass-bottom boat tour out to one of Bermuda's reefs that week, and by chance it was captained by Wingate's son-in-law, who'd said, "Don't even get him started on the Brazil pepper." I didn't need to; Wingate got himself started on it.

Brazil pepper is a many-branched tree with serrated leaves, actually rather lovely, that was imported to Bermuda in the twentieth century. It has become naturalized and is now, according to Wingate, the worst invasive plant on the islands (it is also a severe problem in the Everglades and other parts of south Florida). It readily forms dense thickets that prevent virtually any native species from growing. Starlings are now superabundant in Bermuda's suburbanized landscape—they feed on its lawns and nest in its limestone holes or in the ever-present cavities in its stone houses—and they love Bermuda pepper fruits. They also love to roost on Nonsuch, which, with its high north cliff and no rats, provides plenty of comfortable and safe rock cavities. As a result, starlings are constantly flying to Nonsuch and defecating Brazil pepper seeds all over the place.

"The starlings," Wingate said, "have been an absolute disaster.

The increase in Brazil pepper corresponds exactly with the increase in starlings. The Brazil pepper is like something out of science fiction. I do a hell of a lot of culling. It's my recreation."

Sure enough, as we walked Wingate frequently muttered "There's one" and plunged off the trail into the underbrush, then stooped to tear out a six-inch seedling. Wingate told me later how he fantasizes about training a goat to eat only Brazil pepper seedlings and not those of native plants. He dreams idly about trying to lure starlings from the safety of their roosts with alarm calls on dark and stormy winter nights so that they will be dashed to their deaths in the waves—but how, he wonders, could he ensure that native birds would not come forth too? So for now he teaches plant identification to some of the growing cadre of Bermudian volunteers who like to help out with the project, and he spends uncounted hours stooped over or on hands and knees, pulling weeds and developing an ever-greater familiarity with what he's planted.

It is through this enforced, close observation of the woodland's continuing evolution that Wingate finds some of his greatest joy. Like a proud father, he likes to show off some of his favorite plants. Not species, individuals. One is a wax myrtle. On Bermuda's mainland this shrub, a native, grew only in peat bogs in the 1960s, and so Wingate planted most of his specimens in Nonsuch's wettest places. "I assumed that was its native niche," he said. "But when they matured and started self-seeding they spread all over the uplands. In precolonial Bermuda they were competitive in the uplands. Now that we know that we can refine the project and plant them there. This"—he pointed to a gangly shrub just off the trail—"is the plant that gave me the initial insight." By the same token, the rapid growth of the hackberry trees in the vale told him that they were plants that truly do prefer low, wet places. These lessons could only be learned in the school of hard knocks, but, he said, "If I had to do it over again, I could probably get from Point A to Point B in half the time."

He was proud, too, of a pair of uprooted palmettos that leaned parallel to the ground off to the left of the trail. By the late 1980s, he said, the woodland here had grown dense enough that he wondered how cahows had ever been able to excavate burrows in such a place; there was no bare dirt to be seen. Then a severe hurricane hit and toppled these trees, leaving mounds and furrows of earth that to his eyes looked just like what cahows would look for.

"I await catastrophic events like hurricanes or lightning or pests," he said. "In terms of this nature preserve a hurricane is exciting." It was hurricanes that caused five new native species to sprout on the low sand dunes behind the island's largest beach by clearing space and distributing their seeds. By opening up gaps in the forest, hurricanes also clear space for cedars, which require full light. In the past twenty years Wingate had planted cedars all over the island wherever another tree was blown down. The largest now were overtopping the broad-leafed trees around them.

Best of all were the seedlings. "There are tens of thousands of seedlings under every olivewood tree," he said. "At some point the project developed a momentum on its own that was really sort of freewheeling. Now it's pretty well self-sustaining, but there isn't an end point. What I'm doing is restoring a dynamic process. Any climax community"—or the community that is the eventual end result of ecological succession—"is subject to catastrophic events that set back the clock. And we have to live with some nonnative species that are just too common to get rid of. There are going to be ongoing culling operations. That's the price we have to pay. Restoration is the art of the possible.

"The real enjoyment is seeing how well everything is doing. I get so much satisfaction just walking around the island that culling is no burden."

We walked under the tall hackberries to a wooden blind and looked out from it onto the cattails and open water of the freshwater pond. A dark gray moorhen swam in slow circles and dabbled its clownish orange beak into the floating slick of duckweed. This native species had found Nonsuch, and the pond, on its own. Overhead, kiskadees called, and Wingate spoke of how he'd wanted to get rid of them on Nonsuch but had been unable to do so; they recolonized the place too quickly. Now, to give the native birds a chance, he shoots a few while the vireos are nesting. Yet he seemed sanguine, for the most part, about the nonnative species on the island. He categorizes them pragmatically, according to whether they are a problem or not.

The cane toads, it turned out, were a problem. Large South

American toads that grow almost as large as dinner plates, they'd been introduced to Bermuda to control cockroaches. Some of them had managed to smell the fresh water of this pond from the mainland and had swum across the narrow channel from Cooper's Point. Now they lived on Nonsuch and came to the pond to breed.

Cane toads live in burrows and have highly toxic glands on the back of their necks. Wingate worried that a cahow, if and when one were to return to Nonsuch, might find itself sharing a burrow with a toad. If it were to peck at this unwonted roommate, it would probably be poisoned. "Clearly we've got to get rid of the toads out here before we can invite the cahows back in," he said. "Otherwise we're inviting them into a potential deathtrap." For that reason, Madeiros' crew was at work erecting a three-foot-high black plastic barrier all around the pond. The idea was that toads would head toward the pond and be unable to cross. They would be easy to catch. After a year or two, Wingate and Madeiros figured, all the toads should be gone from the island, and then it would be time to invite the cahows back.

The barrier, it turned out, worked well. By the end of 2003 more than a thousand cane toads had been caught, and the rate of capture was dropping sharply. In the fall of that year Hurricane Fabian struck Bermuda head-on with sustained winds of more than 120 miles an hour and gusts of more than 150. All over Bermuda—which has weathered a great many hurricanes—casuarina trees toppled. The native cedars fared much better. On Nonsuch only a few trees fell, although waves destroyed the saltwater pond. On nearby Southhampton Island a stone fort dating from the sixteenth century was partly demolished by the surf. Driven by the storm surge, thirty-foot waves washed entirely over the small cahow islets. Many of the nesting burrows, both natural and artificial, were destroyed. Madeiros, Wingate, and others worked feverishly to repair them and build new ones before the cahows returned in early winter. Solar-powered speakers broadcasting cahow calls were placed next to the new burrows. In a few cases pairs of cahows that didn't home in on a new burrow on their own were picked up and placed in new burrows. The gruff matchmaking seemed to work, but it was only a temporary solution.

"The storm," Madeiros told me a few weeks later, "underlines that we need to get the birds off the marginal islands where they are now and to islands with deep soil where they can dig their own burrows." Global warming, he and Wingate knew, is likely to result in higher sea levels and ever more ferocious storms. In May and June of 2004 Madeiros and volunteer helpers moved fourteen cahow chicks from their nests to artificial burrows on Nonsuch; in 2005 they moved twenty-one. Fed by hand on squid, anchovies, and sardines, all the chicks fledged within about a month. Madeiros hopes that when the birds return to breed beginning in about 2008, they will recognize particular burrows on the soil-heaped flank of Nonsuch, rather than the craggy islets, as home. In 2006 he'll begin playing cahow calls, at night, on Nonsuch to give returning birds the idea that it can serve as home.

When he took on the job of conservation officer, Madeiros moved his own young family into the island's old house. I was envious of him as I was envious of Wingate's long experience with this place. Visits to Nonsuch are necessarily brief. The island is small, its ecology delicate; there is no place to overnight. What a paradisical place to live, I thought, on a small island laved by clear waves, with the palmetto fronds rustling and the vireos sweetly singing out the name by which they are known on Bermuda: "chick-of-the-village!" I thought of Wingate's young daughters collecting crab parts and trying to catch lizards and watching, from the cliffs, as sea turtles lazed in the patches of eelgrass that grow in the sandy shallows around the island's north face like so many maritime pastures. Whatever the ecological story here, this island was certainly a tremendous place to live and to grow up.

Wingate, too, was a bit envious of Madeiros, who appears as passionately interested in Nonsuch's future as the older man. "I still think of it as my house," he said of the place the Madeiros family now dwelled. "You can imagine, after forty years, it's hard to let go." To some extent he still thought of it as his job too. Wingate had waited to retire until he had both found the right successor and reached mandatory retirement age. Now he was a volunteer. "I'm retired," he said, "but doing the same amount of work as before." Now, in exchange for his volunteer work, he was hoping to be able to live in a second, smaller house on Nonsuch and was half-moved into it, although it had no electricity; he also had a place to stay on the mainland with a friend. He was negotiating with the government for a free lease on the Nonsuch house, and those negotiations were giving him heartburn.

Wingate had always thrived on his independence in conducting his restoration work. He flew mainly under the radar and was appreciated more by international conservation organizations—which continued to fund some of his projects—than by Bermuda's government. "Basically, I probably could have done anything I wanted out here, because they gave me such a free hand," he said. "There was no committee to approve or disapprove. And this really could only have been an individual project. Had it been a committee project it would have ended up looking something like a camel, as they say. But now, to hear the minister of tourism speak, it's the savior of tourism in Bermuda."

Tourism in Bermuda has been in decline for some years, and government officials were hoping that ecotourists interested in seeing cahows and other native animals and plants might revive it. Some calculated that a guesthouse on Nonsuch might produce a tidy income and didn't want Wingate living there for free. The negotiations were tough.

Wingate is well known in Bermuda, and he was planning to pull every string he could to be able to stay. "I wouldn't try if I didn't know that I'm physically able and can handle it," he said. "It's paradisical, but it can be a tough place to live too, physically demanding." He was planning to do the same to try to derail a proposed motor-sports park that a local developer had proposed for the site of the now-abandoned U.S. base on Cooper's Point. He wanted to see the place restored instead—possibly even to cahow habitat. "It's the last chance for an ecotourist project on the island," he said, "and they're going to bugger it up." Clearly, I thought, there was no retiring for this man, not after so many years.

On my final visit to Nonsuch I had some time to myself and walked to the island's west tip, a high bare spot on which Wingate had built a sample artificial nest. He liked to take visitors there. The place had a great view of Castle Harbour, Cooper's Point, the airport, and some of the cahow islets. Waves churned on the reefs far off to the south. Whitewashed hotels and houses clustered densely on the shores of the mainland. Sunlight glared off the choppy water. Tropicbirds and common terns flew by, calling brightly; the former nested in the cliffs just below, the latter on low, flat islets in the sound. Below me I could see Wingate, his boat newly repaired, speeding back to the mainland on an errand.

It was difficult to say what was more thrilling—the thought of the cahows on their nests, hidden from sight now among the darkness of the crenellated rocks as they had been for all those centuries, or flying free above the Gulf Stream's pellucid waters, entirely beyond the ken of human beings for more than three hundred years; or the single-minded dedication of the man who had come along at the precise historical moment to rescue the bird from oblivion, and the resulting and extraordinary depth of experience, the lifetime of observation and hard work and failure and success, that he now carried with him every time he motored under this cliff.

⅜

AFTER WE LEFT NONSUCH THAT DAY, Wingate invited me to go look for a ruff, an Old World shorebird that must have been blown off its northward migration from Africa to Europe and had earlier been spotted at one of Bermuda's golf course ponds by a local birder. We picked up hot mussel pies that we planned to eat at the pond, and Wingate piloted his old Subaru past the "No Trespassing" signs and onto the winding golf cart paths of the Mid-Ocean Golf Course. They were even narrower than most of the sinuous Bermuda roads, and the well-dressed golfers stared in surprise as we puttered past the manicured fairways and stands of wispy casuarinas, no doubt wondering why accidental tourists taking a wrong turn would be driving such a beat-up car.

The pies were cooling and we were both hungry, but first we had to stop in a scrubby woodland opening where the course's maintenance workers dump trimmed vegetation. It was surrounded by lush Chinese fan palms and Brazil pepper trees. Wingate spotted a flash of movement: an indigo bunting, he thought. "It would be about the first spring migrant," he said, as he hopped out of the car. We couldn't find it, though, and Wingate's ears could no longer pick up the bird's high buzzy song. Instead we heard the usual bird sounds of contemporary Bermuda: the cooing of mourning doves, the wheezing of starlings, the maniacal laughter of the kiskadees.

We walked out of the mangroves, through the palms, onto a fairway, and crested a little rise where feral chickens were pecking under casuarinas. Beyond the trees was the pond, and there, sure enough, was a female ruff, a drab brown shorebird pecking at the grass just off the cart path. She was distinguished only by a few feathers that rose up from her back like some kid's Mohawk haircut, which according to Wingate clinched the identification. He pulled out his pocket notebook and jotted down a note, the way he'd been doing for five decades. He noted the ruff, the snowy egret stalking the shallows, the five moorhens picking at the grass, the fledgling mallards lined up behind their parents, the eastern bluebird flitting around the reedy pond edges. Not until he'd gotten it all down did he take a break. With golf balls whizzing overhead, we dug into lunch. Wingate gestured at the pond, which was perhaps an acre in size. All around us the slopes rose manicured in immaculate grass. The golf carts buzzed. "For every Nonsuch Island we've created we're making a hell of a mess out of the rest of Bermuda," he said. "With all the human use and disturbance there's nothing but these pathetic pocket handkerchief bits of nature left."

It was all too true, but I also realized that its intensive development meant that Bermuda's mainland was the ideal place to consider the lessons of Nonsuch. What did it mean here that Wingate had spent forty years restoring a tiny island? Poor Bermuda, I thought, so heavily altered by humans: there was no going back to the enchanted paradise of cahows and terns and sea turtles that Sir George Somers and his men found. On a small island, the effects of humanity were so quickly and so profoundly apparent: Bermuda was a microcosm, depleted of its riches by the malice or ignorance of its new European residents over only decades rather than centuries, as North America as a whole has been. Wingate had just told me that far fewer warblers showed up in Bermuda during the spring migration than had done so when he was a young man, a change due to environmental changes in the Americas rather than in little Bermuda.

But if Bermuda was a microcosm of the continent to the west, then Nonsuch was a microcosm of Bermuda, and some of the lessons Wingate had learned during his long ecological apprenticeship were spreading outward too. People had begun planting cedars and olive-wood trees in their gardens. Ever-increasing numbers of Bermudians are volunteering their time to pull weeds or conduct surveys on Nonsuch. The local Audubon Society is swelling. A new birding guide had just been published. In 2004 the cahow was declared Bermuda's national bird, supplanting the more common and widespread tropicbird. I recalled Wingate's telling me that he didn't think of Nonsuch as a place apart. "The whole concept that you have nature preserves and just shut people out is not going to work," he'd said. "We should treat the whole planet as a nature preserve in which we are just one of the species living in balance."

He meant in part that there is always going to be a need for ongoing maintenance on Nonsuch in particular: people are always going to need to pull the Brazil pepper seedlings. People are doing that and more in many places, as we'll see: tending wild gardens, helping rare species, setting things right. They're practicing restoration—ecological healing—to bring nature back to health.

Wingate, however, also meant something more. He was talking about his own joy in pulling up all those seedlings and about how he always learned more about the island and his own history there during the course of that work. It was a meditation. So when he returned to Nonsuch, each and every time, he had a feeling not of being oppressed by the ongoing and ever-present need to do more work but rather of something else: "I never take it for granted," he said. "Every time it's an overwhelming feeling of joy: 'God, there's presettlement Bermuda. It's back.'"

INTRODUCTION

Prospero's Task

One can scarcely think of David Wingate, with his sturdy physique and his grandfatherly mien, as other than the Prospero of Nonsuch Island, the magician with a heart of gold who through singular dedication sets things right. That's how it goes in *The Tempest*, after all. Prospero, banished from his dukedom by his avaricious brother, works his good-natured sorcery and the spirits of an enchanted island to right the wrongs of the past. In the end, restored to his proper place in the world, he is able to leave what he calls the "bare island" where he was imprisoned. The slate is wiped clean; the miracle worker departs.

When Wingate says, "God, there's presettlement Bermuda. It's back," then, it is tempting to take his words at face value. Because of his obvious successes and evident determination, it is tempting to believe that he's indeed accomplished everything he set out to do and that now he can step out of the picture, to applause, and let Nonsuch Island go on by itself.

It *is* tempting. The human heart yearns to wipe away the mistakes of the past, and nowhere is that tendency stronger than in America, land of the fresh start. But it is an impossible ideal. Leave aside for the moment that this particular Prospero has not been making a very clean exit from the place of his labors. In addition, presettlement Bermuda isn't back on Nonsuch, and, with his aching back, no one knows that better than David Wingate. It is hard work stooping to pull all those Brazil pepper seedlings, and however fond his hopes he knows that they aren't going away anytime soon. Neither are the starlings that scatter the pepper trees' seeds all over the place, or the kiskadees, or the cardinals, or the anoles, or many of the other non-native species now at home on Nonsuch. They're there to stay, altering the island's ecological circumstances in ways no one in Juan de Bermúdez' day could have imagined, and in ways that will probably continue to surprise ecologists in the future.

Wingate knows intimately that it is going to take the dedication of people like himself, and like Jeremy Madeiros, not just for a few years or decades, but in perpetuity, to keep Nonsuch Island in cedars and wax myrtles and cahows. It is going to take weeding and other sorts of manipulations, year after year, and it is going to require the sort of extended human caring and dedication that can seem a mighty slender reed in a storm of modernity. It is going to take paying attention and dealing with surprises: who could have expected that keeping out cane toads would prove important in allowing cahows to nest? Without active human intervention, generation after generation, Nonsuch will eventually slip back into the sort of weedy, non-native vegetation that covers much of the rest of Bermuda. In the real world Prospero simply doesn't get to leave.

True, on Nonsuch Island one *can* get a very good idea of what Bermuda once looked like, and thanks to Wingate's and Madeiros' hard work the likeness to an earlier time may yet get better and better. The cahow has clearly been helped past the yawning abyss of extinction. Perhaps the sea turtles will come back to nest too. The entire ecosystem will likely persist in greater health than on the rest of Bermuda: it is clearly far more resistant to the inevitable hurricanes, for example, its native plants ready to fill gaps in the woodland canopy when a few trees do fall. Yet what is going on there is not—cannot be—quite the same thing as returning to the past, or even bringing the past into the present.

⁂

IT IS, INSTEAD, SOMETHING NEW. The work on Nonsuch Island is one of the premier examples of ecological restoration, a burgeoning practice that is redefining the place of people in nature. Restoration is ecological medicine. It proceeds from the premise that people have done a great deal of damage to our natural surroundings and that it is therefore a human obligation to reverse that damage as much as possible. Sometimes the damage is deliberate, as with the early slaughter of the cahows. Often it is quite accidental, as with most of the nonnative plants and animals of Nonsuch Island, most of which were imported for what seemed like good reasons at the time. In either case the David Wingates of the world—and there are many— have decided that it is time to intervene, to use human energy and ingenuity to alter these places in a positive way.

Their techniques are as manifold as the practices of medicine. Sometimes healing proceeds from the simple act of stopping something harmful. A doctor tells a patient to quit smoking; a law forbids the killing of cahows. Doctors, however, know that is often not enough. They rely in large part on more intrusive practices—surgery, precisely tailored pharmaceuticals—and so does ecological restoration. Wingate's decades of experience on Nonsuch have showed, time and again, that simply walking away and expecting the island—or the cahow population—to heal on its own is not possible. Active intervention is required, and it can be as simultaneously heavy-handed and as delicate as open-heart surgery. If the soil is eroding, you're going to have to plant something. If you want native plants to grow and their seeds aren't being produced locally, you're going to have to plant them yourself. If weeds threaten to outgrow the native plants, it may be necessary to get on your knees and pull them out. If cahows aren't coming back on their own, you might try constructing artificial burrows and playing tape recordings to bring them back—after first getting rid of the cane toads. And so on. Bermuda, so small and so far out at sea, is ecologically a lot less complicated than larger islands or continents, but David Wingate's long experience reveals just how difficult it is to restore the vitality of even a simple ecosystem.

This book is a tour of the new landscape of restoration. It is about the efforts of a series of Prosperos who have spent years and decades and entire lifetimes practicing ecological medicine. In some cases they're motivated, as Wingate initially was, by the desire to help a single plant or animal out of a tight spot. In many others they're focused on entire ecosystems and want to ensure that complex suites of plants and animals can continue to coexist together into the future. In quite a few instances they're motivated as much by a desire to improve human welfare—economic, aesthetic, or spiritual—by repairing broken relationships between people and their surroundings.

I am going to make a sweeping claim here: restoration is going to

become one of the dominant ecological and social movements of the
twenty-first century. It simply has to. The simplest support for my
claim is the old Will Rogers bromide: "Buy land. They ain't making
any more of the stuff." That's abundantly clear in a small place like
Bermuda, where the English settlers learned early on that the paradisical abundance they found was far from limitless. Cahows and
cedar trees and reef fish all seemed inexhaustible resources at first.
When exploited, though, they pretty quickly showed themselves to
be, indeed, "only skin deep," as subject to overuse and wearing out
as the long-pressed wildlife and trees and fisheries of England. Simply using up what the land provided, with no thought for tomorrow,
was a strategy that only worked for a few years in Bermuda.

It worked a good deal longer in the vast expanses of the world's
continents, but even here limits have become painfully obvious. Consider a few statistics culled from a recent United Nations assessment
of the world environment. Since the end of World War II more undeveloped land has been converted to cropland than had been in the
entire eighteenth and nineteenth centuries. Since 1960 the amount of
water impounded behind dams—and lost to rivers—has grown four-
fold. Overall human water use has doubled, wood harvests have
more than doubled, and food production has increased two and a
half times. All this human use of land and resources has already
increased the extinction rate for wild plants and animals as much as
a thousand times over prehistoric levels, and about a quarter of the
world's remaining animal species are threatened with extinction. It
has also increased stresses on many natural processes that make
human life possible, such as pollination, erosion prevention, the
cycling of clean water and clean air, and provision of food and fiber.
Two-thirds of these vital services, according to the United Nations,
are in decline, making it ever more difficult to sustain human liveli-
hoods and economies, especially as human populations continue to
increase.

North America has no exemption from these trends. In the United
States, roughly a thousand species of plants and animals are officially
ranked as threatened or endangered and at risk of extinction, and
conservationists claim that many more ought to be listed. One-sixth
of North America's area suffers from severe land degradation.

Increasingly severe wildfires ravage western forests. Soil erosion is rampant. Invasive plants are simplifying what were once complex ecosystems. Sprawl and road building threaten both the continued survival of many plants and animals and human enjoyment of natural landscapes. The bill for this ongoing neglect of our surroundings is large: in 2005 Hurricane Katrina's devastation of the Gulf Coast was simply the latest graphic reminder of what can happen when ecological processes—in this case, the buffering of floods provided by coastal wetlands—break down through carelessness or outright abuse.

Given these trends, there can be no disputing that restoring degraded lands and waters has got to be one of the central human activities of the twenty-first century. There is simply no more virgin land for the taking. The idea that we can mess up a place and move on, so central to U.S. history, has far outlived whatever usefulness it had. It is time for something new, say the modern-day Prosperos: time to fix what's broken, even if it's badly broken; time to clean up our mess and learn how to live so that we don't make too many new ones. That is the central premise of ecological restoration.

It is an idea that is taking hold all over the world, including all over North America. In any good-sized city and in many smaller communities as well you can find groups of people who have made it their business—generally unpaid—to repair streams, marshes, woodlands, prairies. They're fixing Prospect Park in Brooklyn and the Anacostia River estuary in Washington, D.C. They're revegetating moonscapes of nickel-mining slag around Sudbury, Ontario. They're burning myriad tallgrass prairie tracts throughout the Midwest. They're rehabilitating the dunes of San Francisco's Presidio and salmon streams throughout the Pacific Northwest. They're bringing native forests back to the lowlands of the Hawaiian Islands.

They have also, in many places, prevailed upon governments to carry out the same sort of work at a larger scale. In the 1990s New York City kicked off a $1.5 billion project to restore its watershed in the Catskills—which seems like a lot of money until you learn that it would have cost the city about four times as much to build a new water treatment plant, with additional expenditures for yearly maintenance. In 2002 Congress authorized the expenditure of $19 billion to clean up Chesapeake Bay, where a valuable oyster fishery, along with many other resources, has been declining due to pollution, waterfront development, and other problems throughout its watershed. A similarly scaled project is underway in the Florida Everglades. In July 2004 state and federal officials began the restoration of 16,500 acres of salt evaporation ponds strung around the perimeter of San Francisco Bay.

Projects of this scale matter not only to governmental entities but also to entrepreneurs and to businesses of all sizes. "The restoration of our natural and built environments has become the greatest business frontier of the twenty-first century," writes a businessman named Storm Cunningham in a cheerleading 2002 book titled *The Restoration Economy: The Greatest New Growth Frontier*. Such projects are thoroughly in line with a venerable American tradition— namely, thinking big. You can, if you wish, measure their success by counting dollars and acres and numbers of government agencies involved.

But you can also measure the spread of the restoration idea by paying heed to the devotions of the many thousands of people who spend their weekdays or weekends working to restore patches of forest or marsh or prairie that are often the size of Nonsuch Island or smaller. You might measure it by looking at the motivations of David Wingate and his many far-flung colleagues, and at what pleasure or virtue or challenge they find in their work. This book is an effort at such assessment.

⚘

MUCH OF THE NEWS that reaches people today about the natural world—whether through the news media or through the senses—is bad, and on its face it is not difficult to perceive an overarching sense of environmental malaise among many Americans. In media accounts it is difficult to avoid the impression that anything human beings do is harmful to nature. We cut trees and dam rivers. We strip the productivity from soils and the fish from the seas. We slice remaining wild places into ever-smaller parcels and pump climate-altering chemicals into the atmosphere. We are, ecologists tell us, facing the sixth great extinction event in the long history of Earth, an orgy of loss on a par with the disappearance of the dinosaurs.

All these things are true, yet a compelling argument can be made that dwelling on them is far from an exercise of broad vision. By constantly raising alarms, environmentalists—considerably aided and abetted by media interests that seek sensational stories and controversy—have fed the idea that people can do only harm in the world. That's both a difficult way to motivate people in the long run and a notion that ideological opponents can easily parody for their own purposes. *Environmentalists worry about everything*, the argument goes, *and they're just one special-interest group among many*. As a result, good points about serious matters—what could have a greater effect on everyone than climate change?—go disregarded. The middle ground erodes. A large part of the populace willingly shuts itself out of debates about the lands and waters around us simply because it is too tiring to maintain a sense of caring in the face of so much bad news.

Ecological restoration can break through this public logjam precisely because it presents us with a means of doing good in the world. *Doing*, that is, not in the sense of opposing something bad, but in the deeper sense of acting in a positive way. *Doing*, in the sense that people prefer the word *yes* to the word *no*. As much as it has been a chronicle of taking, the story of the United States has been a saga of building things—farms, cities, dams, highways, cathedrals, spacecraft—and if some of those things have been built at grave cost to the natural world, and to people, it is important to remember that some of them are also beautiful and that many people found great satisfaction in their construction.

Conservationists have proven very good at saying *no*. Some of their greatest victories have come in opposing new dams, highways, and oil wells. But *no* is not a viable political strategy. It is not an effective way to motivate public support in the long run. "Leave No Trace" may be an appropriate way to think about visits to a certain wilderness area, but it is no way to run a planet. Restoration is the best way to maintain the integrity of the natural world, and it is becoming a powerful social movement, precisely because it offers a chance to use human energies in ways as deeply satisfying as building a dam or a cathedral. It perfectly links individual action with the needs of something greater, both in human society and in nature.

Restoration also perfectly weds the intellectual and the physical. As David Wingate came to learn many years ago, it is work—hard work. You come home at the end of the day bone-tired. Your arms have been scraped open by thorns or sharp-edged rocks, your neck is sunburned, and your lower back aches from bending over. The tiredness feels good. Your physical body has been conjoined with your beliefs about what the world should look like. You've made a difference. When you go back in the morning, or the next weekend, you can see what you've accomplished.

Up to a point, that is—for you also realize, viscerally, that there is a limit to what you and anyone else can do. David Wingate learned quickly that it was the cedars and other native plants, and the night herons, that would do most of the work of restoration. His job was to help them along a bit, to give nature a nudge in the right direction. Restoration, at some point, involves letting go. It's not gardening. Restoration does not shape the landscape into something conceived of solely by people. Instead, it hews more closely to the practice of medicine, which can work only if the doctor can provoke the patient's own body into doing the bulk of the healing.

What restoration offers us, then, is a way to use our uniquely human qualities of caring and intellect in playing a positive role in the physical environment. It puts the lie to the popular but unhelpful idea that what's good for people is bad for nature and vice versa. Restoration is a bridge between what has come to be viewed as the disparate realms of humans and of nature. Precisely because it is a bridge, it offers possibilities that don't exist on either of the two banks. Restoration is not just people working, and it's not just nature. It doesn't fit easily into our preconceived categories for organizing the world.

You could use that as an excuse for avoiding the whole affair; it's just too much trouble. On the other hand, you could view it as a new opportunity, rich with possibility and complexity: a new covenant between people and nature.

One of the real possibilities inherent in the practice of restoration is that the myriad Nonsuch Islands of the world will become healthier places that are better homes for cahows—and that retain the many essential ecological services that make human life possible in the first place. Another good possibility is that restoration might just make life more meaningful for the many Prosperos out there, helping them discover new depths in their surroundings, in other people, and in themselves.

࿊

I FIRST HEARD OF ECOLOGICAL RESTORATION when I was a teenager growing up in a Chicago suburb. A small group of people, I learned, was working to reclaim a few patches of prairie that had managed to avoid being turned into housing developments—although they hadn't quite avoided turning into woods that bore little resemblance to prairies. It sounded like a good idea to me. Not until many years later did I—or most anyone else, for that matter—hear that not everyone agreed. Some people didn't think prairies were worth worrying about, at least not if having them around meant that you had to cut trees down.

Years later, as an employee of a university institute that studies forest restoration, I came to edit a book about the restoration of ponderosa pine forests in the southwestern states. In northern Arizona, too, people who talked about restoration wanted to fell a lot of trees. To reduce the danger of wildfires and improve ecological health, they proposed restoring the area's now-dense forests to something resembling the more open condition they were in when the first Euro-American settlers got there. The cutting of trees has a decidedly checkered history, and so quite a few people thought the chainsawing a bad idea. The arguments were heated.

I came to realize, though, that the disagreements weren't so much about ecology as they were about values and about what sort of interactions should exist between people and their surroundings. Restoration, I learned, is one of the great stories of our time not only because of its impact on nature, but because of the myriad issues it forces people to consider. As I'll show in the chapters to come, it touches on all aspects of human life, from the political to the economic to the profoundly spiritual. Restoration is a tale of people who take seriously the idea that human beings should steward the natural world without turning it into a garden crafted solely for human purposes. It is an effort to put into practice Aldo Leopold's admonition that *Homo sapiens* ought to act as a "plain member and citizen" of the land community, rather than its conqueror.

Restoration can also be a profoundly democratic act. Many of those most involved in this work came to it not through professional responsibility or through political appointment, but simply through love. They came to care about a place and, in caring about it, felt obliged to work toward its betterment. Of course, precisely what "betterment" means in talking about natural places can be as contentious as public debates about war, education, or the intersection of church and state. Equally well-meaning people can possess very different opinions. Those debates are part of the restoration story, and they are only going to become more pronounced as the practice grows.

What restoration amounts to is coming to feel truly at home somewhere on the surface of the planet, rather than in some imagined paradise, whether theological or technological. Remember, we've seen that Prospero has to continue working on his island. He may have some benevolent magic, but he also has to come to an accommodation with his place. He has to stay involved, in place. So do his descendants. They're not going anywhere, because there is nowhere else to go. They're staying home. They might have differing visions of what that means; disagreements about what home should look like can become pretty heated. Yet there is reason to hope that those debates can, as they ought to in a democratic society, move everyone's ideas forward.

The stories in this book are tales of a vibrant democracy of the land in action. They are profiles of people who are trying to restore individual animal or plant species, entire ecosystems, even cultural practices that govern how people and nature ought to interact. They're shaping the future for nature and its people. This book is a record of my journey in tracing the history of restoration, visiting its frontiers, and guessing at its potential.

In exploring the course of healing it made sense to start in the east and head west, following the course of empire and exploitation. From Bermuda, where the English landed when the idea of the New World was young, I headed for the green tangles of the eastern mountains.

1

Smoldering at the Roots

*F*red Hebard was practically giddy in the rain. He had a gleam in his eye familiar to anyone who's seen a fourth-grader on a snow day or, for that matter, a farmer on holiday. He wasn't exactly getting away from his work, but he was playing hooky nonetheless.

The hamlet of Meadowview, where Hebard works and lives with his family, is an eyeblink along Interstate 81 in southwestern Virginia. If you were for some reason to exit there, as few people do, and wind up at the Little Diner for lunch on a sultry June day, you'd probably mistake Hebard for one of the cattle or tobacco farmers who have, over the course of more than two centuries of husbandry, made the fertile Holstun River valley a prosperous agricultural area. He'd probably be wearing a T-shirt and jeans and, atop stooped shoulders and graying hair, a baseball cap reading "Big M Farm Services." He might well be flashing a wry grin and joshing about cars or the weather with the other farmers and the tractor mechanics. Then, after a Little Burger—which isn't so small at all—and a smoke, he'd rush back, like the others, to the endless whirl of work that is a farm in June.

This is just some of what occupied Fred Hebard on that June morning before the rain: all day long he had to oversee the endless details of running three farms, totaling more than a hundred acres in size, as the thick air coalesced into hazy thunderheads. He had to ensure that three full-time employees, two seasonal workers, a summer intern, and a dozen Elderhostel volunteers who were staying at a nearby 4-H center and dedicating their week to his projects all had timely chores to do, the training to accomplish them properly, and the appropriate equipment, from a backhoe to a bucket truck to arrays of petri dishes inoculated with fungus. The bucket truck was of the variety used by professional tree trimmers, but it was an older model, and the farm-hands were acutely aware that it smelled as though something had crawled into its hydraulics and died. He had to walk among rows of chestnut saplings, twice as tall as he was, and assess which ones deserved to continue growing, which ones should be pollinated by hand in coming weeks, and which ones would be ripped out that same day with the backhoe. Above all, he had to carry in his mind, as he has done every day, year in and year out, over a career of twenty years, a mental picture of the some 17,000 chestnut trees he has raised, of what needs to be done to stimulate their growth, and of which ones thrive and which do not.

It was easy to imagine where the stooped shoulders had come from, and so it was not hard to imagine why the gleam grew in his eyes as he watched the clouds form and the storm approach. He told the Elderhostelers to stop working because of the danger of light-ning, and then half an hour later, as the rain began to fall in earnest, told Dave Lazor and me that he had something to show us.

Lazor is a retired General Motors engineer from Ohio who had spent his career figuring out how to get the manufacturer's machine tools to run more efficiently. He is also a forest enthusiast who'd recently bought a hundred-acre woodlot in western Pennsylvania. When he did he recalled how someone had given his father a gift of some beautiful chestnut boards in the 1950s, back in the day when it was thought that no more chestnut lumber would be forthcoming in the future, ever. He soon learned that American chestnut trees, in fact, grew right there on his woodlot, and since that discovery he had brought some of his engineer's attention to efficiency to forestry. His passion became figuring out how better to grow chestnuts.

For Lazor, this visit to Meadowview was like a Catholic's trip to Rome. Fred Hebard is the staff pathologist, head farmer, and chief petri dish washer for the American Chestnut Foundation, which has the goal of returning an almost-vanished species, once a hallmark of

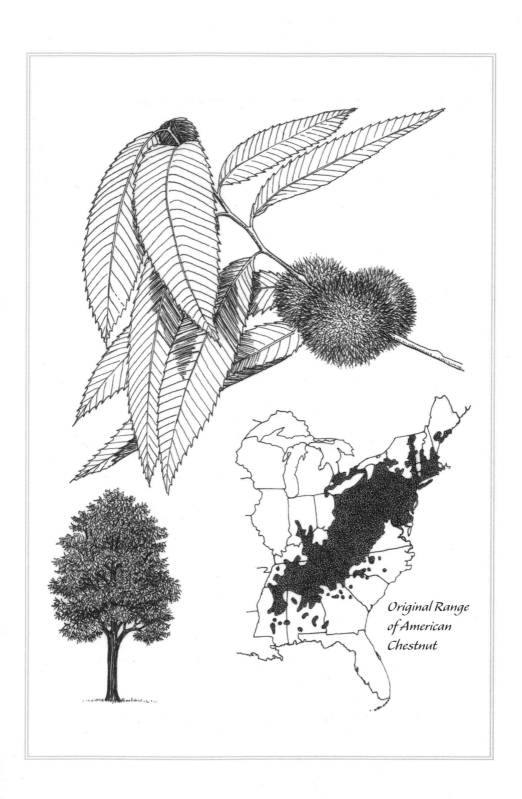

*Original Range
of American
Chestnut*

the eastern mountains, to its place in the landscape. If anyone can bring the tree back from the brink, Lazor believes, it is Fred Hebard.

෨

THE WOODS, AS IT TURNED OUT, were on the high ridge to the south, part of the Jefferson National Forest. As we climbed slowly out of the valley a last couple of houses stood by themselves in a misty clearing, and then the woodlots and isolated groves turned into a solid wall of trees. Lazor pulled his pickup truck up to a locked gate. "Here's one hundred thousand acres of prime chestnut habitat, and I've got the key to it, ha, *ha*," chortled Hebard as he stepped out to open the gate. He's got a peculiar laugh, with an emphasis on the second *ha*, and he used it often on this outing.

The woods were dense and dripping. There were a few pines and hemlocks, but most of the trees were deciduous: oak, maple, tulip poplar, sourwood, sassafras, cherry. Lazor wound the pickup slowly around the sharp bends. The narrow, rain-slicked road was built for logging, not public use. "Just don't roll the truck," Hebard said. "I've tried that experiment and don't want to repeat it." When he had slid his truck over a forest embankment, he told us, it had done a full 360 and came to land back on its wheels. "If it had landed on the roof," he reflected, "I probably would have been in a bit rougher shape. Ha, *ha*."

We trailed a few miles up into the mountains until Hebard told Lazor to stop in a small, grassy clearing. This tract of land had been clear-cut eight years earlier to serve as the landing where the hardwood logs were loaded onto the trucks. Southwestern Virginia is a lush, moist place, and to my untrained eye the place scarcely looked like a former clear-cut: tall grasses glistened on the roadbed, but all around them trees reached thirty feet high. The rain was steady now, and soon we were soaked. We were five hundred feet higher than in Meadowview, and the air was noticeably cooler. Mist hung in the treetops just above us. Hebard gestured at the trees on the uphill side of the clearing. "There they are," he said.

We looked, and there among the crowns of the commonplace maples and tulip poplars was an extraordinary sight: sprays of creamy white flowers shooting up above the highest leaves. It looked

as though, in the midst of this gloomy day, a few of the trees had captured filmy highlights of sunshine high in their foliage. The trees were
American chestnuts, and for a species presumed extinct by many they
were putting on a pretty good show.

Their performance was not entirely natural. Every summer since
the logging Hebard had driven up here with his helpers to chop back
the competing brush and give the chestnuts a better chance to grow.
Every summer since the chestnuts began blooming he'd driven the
bucket truck up here and parked it precariously on the steep road so
that the blossoms could be laboriously pollinated by hand. Each
autumn he had returned to collect the nuts before they were all eaten
or carried off by squirrels, turkeys, and other animals.

Hebard had, in other words, come to know these trees as individuals. He pointed to an unruly clump of trunks to introduce us to a few.

The trunks were five and six inches in diameter, and they were
all marred by long, vertical strips of scar tissue that rose like long
Appalachian ridges through the otherwise smooth, reddish brown
bark. They would cause most home gardeners to wince, but to
Hebard they were signs of resilience, an indicator that these trees
were resisting the fungus by building scar tissue to shield their vital
sapwood. They were fighting. To Hebard, whose entire career has
focused on conflict at the microscopic level, it is such cankers that
represent the future of the American chestnut.

"Can you see the big ugly cankers here?" he asked. "That's the technical term, 'big ugly canker,' ha, *ha*. Sometimes when there's fair resistance you'll get trunks that are all big ugly cankers, fifteen feet high."

The trees may have represented the future, but even to a nonex-
pert like me it was pitifully clear that they didn't have much of one as
living individuals. Their leaves were small and sparse, and the neighboring poplars and maples were growing higher and faster than they
were. Hebard's ministrations had allowed them to grow and reproduce, but these chestnuts were dying as so many others of their kind
have. Within a few years the canopy of other trees would close in and
the chestnuts would be little more than stark wood skeletons that
would soon fall unnoticed and through their decay enrich the soil—
and Hebard would look elsewhere for chestnuts able to bloom and
put their energy into future generations.

He is unsentimental about this, as anyone must be who is in the business of farming genes for the future. Only through the deaths of uncounted thousands of trees that don't have the best genes for survival against the blight can he find those that do. Hebard indeed seems an entirely unsentimental man, but as we walked back toward the truck he suggested we take a short stroll down a winding side road that opened off the clearing. The rain showed no sign of letting up. Water ran in rivulets through the grasses and sedges and mosses. My shoes were drenched. Somehow Hebard managed to keep a cigarette lit. We rounded a few turns in the road and came to an open grove where several tall tulip poplars two feet in diameter grew in front of a little pond. A few white pine saplings, as tall as a man, grew among the grasses and wildflowers.

"I don't know if it was worth getting this wet," Hebard said, "but I just really love this glade."

There wasn't a chestnut tree in sight, and I realized he'd brought us here for another reason, simply because he liked the place—and because it helped fuel his work. He looked at the tall poplars and the sapling pines, and reflected on why these particular trees were growing in this particular place.

"So many of these forest changes," he mused, "take place beyond the memory of any one person, so they can be hard to understand. Of course you can take a look at stands in different stages of succession and begin to interpolate, but it's so complex." There was that glint in his eyes. He flashed a boyish grin, and I was struck by the realization that he loved the complexity of ecology, even as it made his job harder. He was a farmer with sizable and clearly laid out responsibilities for breeding and growing trees down on the farms in the valley, but his heart was here, in the tangled woods where the chains of cause and effect are so very hard to tease out. His obligations did not end at the edge of the cultivated fields. Rather, they were to the forest itself.

Perhaps he is perfectly disguised as a farmer, but Hebard's goals are entirely unlike those of conventional farming, which aims to corral nature for human purposes. He, along with a number of people of similar dedication, are attempting something much more audacious, a sort of farming in reverse. They are using the methods of farmers—

the tractors and herbicides and fertilizers and means of tracking genetics—to create a cultivar not so that it can be more effectively used by people, but rather so that it can be returned to the wild. It's as if a farmer were to develop a field of corn with the aim of having it take care of itself for hundreds of years, without any further human intervention. If Hebard and others succeed—and whether or not they will remains very much an open question—they will not only return a once-vital plant to its landscape, but also extend the frontiers of what people can do to set right the mistakes of the past.

ॐ

WHEN THE CHESTNUTS DIED, the ghostly trunks of the dead trees cast such a pall on the landscape that some thought "the whole world was going to die," as an eastern Kentucky resident recalled much later. It was, as one elderly Tennessean told an oral history interviewer in 1965, "the worst thing that ever happened in this country." He was speaking of the country in and around Great Smoky Mountains National Park, but he really could have meant a far larger area than that: chestnuts grew from the southernmost extensions of the Appalachian Mountains, in northern Alabama and Georgia, west across Tennessee and Kentucky and Ohio, and as far north as Maine and Ontario. Throughout much of that huge range the chestnut was the dominant forest tree. A member of Hernando de Soto's 1540 expedition through the Southeast put it in the simplest terms possible: "Where there are mountains, there are chestnuts." Two hundred million acres of chestnut forest were splashed across eastern North America. Chestnuts shared the slopes of Stone Mountain in Georgia with magnolias and laurels and hickories. They grew on Lookout Mountain in Tennessee and towered alongside oaks and basswoods throughout the heightening Alleghenies of Pennsylvania. On the hilltops of New Hampshire they grew, prodigiously, next to beeches and maples and cherries. They were catholic, weedy, tolerant of elevations from sea level to five thousand feet. Across much of the landscape they made up a quarter of all the trees; on the Highlands Plateau of western North Carolina they were half of all the hardwoods. In some places nearly pure stands of chestnuts stretched for a hundred acres and more.

Imagine chestnut trees as you might imagine maples today in New England or spruces in Alaska: as simply ubiquitous. Imagine asking a child in Appalachia, a century ago, to draw you a tree: chances are a chestnut would have served as the model. The trunk would likely have been long and straight and thick and covered with pale, deep-fissured bark. Chestnuts sometimes grew more than a hundred feet tall. Henry David Thoreau, on his home ground in Concord, Massachusetts, measured a fresh chestnut stump twenty-three feet and nine inches in circumference. In the southern Appalachians, where the growing season is longer, a chestnut tree grew to seventeen feet in diameter in Francis Cove, North Carolina. In the woods the trunks of mature trees might shoot up fifty feet or more before any limbs branched off, but if it grew in a field, a chestnut could sprawl magnificently, its dense foliage casting a wide pool of shade over kith and kine.

A leaf in the drawing would have roughly the shape of an ash leaf, curving elegantly outward and then tapering to a point, its two sides like surfaces of a lake ruffled to a sawtooth chop by a fresh breeze. And if it were June or early July, chances are the artist would draw chestnut flowers too—six- to eight-inch-long spikes of tiny cream-colored blossoms, clustered together in starburst formations reminiscent of Independence Day fireworks and all jangling together in the breeze so that the pollen could be blown from bloom to bloom, which grew so dense sometimes on hillsides that it looked as though it had snowed. If it were October instead, no child would pass up the chance to draw the single most essential part of the tree: the nut, a smooth-skinned, velvety brown treasure tucked inside a ferociously prickly burr like a wrapped present, perfect for eating, but also for throwing or fondling restlessly in a pocket through the dull days of winter.

There were other trees, of course, especially in the diverse forests of southern Appalachia where chestnuts reached their greatest size and abundance. White pine grew taller, and oak made more durable furniture. Hickory made better ax handles and wagon wheels. Maple produced its unique syrup and burned longer and hotter in the woodstove. Cherry and walnut, or such imported rarities as mahogany, were the only woods really adequate for the furniture and paneling of a rich man's study, but only chestnut combined so many useful virtues in a single accessible package. It was everyman's tree. It was the singular natural resource that fueled the first surge of American manifest destiny through the Cumberland Gap and across the Appalachian chain. Like the Scots-Irish pioneers who settled the eastern mountains, it was tough, weedy, aggressive. It grew and thrived almost anywhere. It was sturdy and unpretentious and could be turned to a hundred uses. If any single organism in the United States could represent the ideals of Jeffersonian yeoman democracy, the chestnut was it.

No tree could have been designed to meet more efficiently the needs of animals and people. For starters, chestnuts grew fast. A chestnut could grow to the height of a tall man in a year. In twenty years it could grow halfway into the forest canopy and make an ideal telegraph pole, straight and free of knots. If it were cut, numerous sprouts sprang up from the roots, within a year forming, as Thoreau wrote, "perfect bowers in which a man might be concealed." A tree that wasn't cut could grow to more than four feet in diameter in the span of a single human lifetime and could survive for six hundred years.

A chestnut forest could produce prodigious quantities of lumber. In Scottsville, Kentucky, seven hundred fence rails were split from the trunk of a giant tree, nine feet in diameter, that was felled in 1876. Settlers built entire houses from the wood of single trees. The wood was easy to work; old-timers said it "split like butter" when it was green. Like the wiry mountaineers themselves, however, chestnut logs grew tougher with age. They were resistant to decay and to the workings of most insects, especially when set into the ground. Chestnut saplings, charred at the tip in an open fire, made splendid fence posts, and throughout Appalachia you can still today see the winding "snake fences" made of split chestnut rails. There's one, salvaged from the Blue Ridge, in front of one of the farms Fred Hebard manages in Meadowview. Sheets of chestnut wood were light and didn't warp. They took glue well and made an ideal base for veneer; much old American furniture ostensibly of oak or walnut is really chestnut with a thin veneer of the more expensive wood. People throughout the mountain region literally could go through life and death surrounded by items built of chestnut; it formed cradles and caskets, boxes and barrels, pallets and pianos, logging skids and railroad ties. Boys made whistles out of chestnut bark, and hopeful young lovers inscribed their names on the smooth bark of saplings. Chestnut firewood heated many a woodstove; it caught fast and, like many young loves, burned hot if not long. It maintained an even temperature and for this reason was a preferred fuel for the process of annealing brass. Where sugar maples grew, the chestnut was often chosen to fuel the springtime ritual of boiling sap into syrup. In Pennsylvania and the Northeast charcoal producers favored chestnut for its habits of sprouting and quick growth as they felled massive numbers of trees to feed the growing nation's iron foundries. The bark, meanwhile, was stripped off to make tannin for the tanning industry, which by the early twentieth century used more chestnuts than trees of any other species. Railroad cars full of hides from the Chicago stockyards were shipped to southern Appalachia so that they could be tanned there.

Pioneer families, like the Cherokees and other native peoples before them, found other uses for the tree. A tea steeped from its leaves served as a cough syrup; leaf poultices soothed the pain of burns and poison ivy rashes. Mattresses could be stuffed with dried leaves. Bees made excellent honey from the abundant blossoms. When the blossom tassels dried and fell off in summer, boys collected them and rolled them up in paper to make potent homemade cigarettes.

Had it borne no crop at all, then, the chestnut would have been an important resource to people living in and near eastern forests. Yet it had a crop, an extravagant one. Most eastern forest trees bloom in spring, when late frosts can destroy the developing nuts or fruits. Chestnuts wait until early summer, when the danger of frost is almost entirely past. As a result, chestnut crops were reliable, year in and year out. A big tree could bear six thousand nuts, all carefully encapsulated in astonishingly prickly burrs. Some boys dared one another to stomp on the burrs with summer-calloused feet to expose the nuts hidden away inside; for most country children the dropping of the burrs was a visceral signal that it was time to put on shoes again. Hunters told tales of bears that wielded rocks to smash still-closed burrs to get at the nuts. Chestnuts have high protein content, some fat, and a fair amount of sugar. Nuts borne by American chestnuts are generally smaller, but also sweeter, than the commonly available Asian and European varieties. They made fine feed for animals wild and domestic. Bears fattened up on them in preparation for hibernation. Squirrels made great caches of them. A hunter in what was to become Great Smoky Mountains National Park shot a wild turkey whose craw held ninety-two undigested chestnuts. Before they became scarce and then died out altogether, great flocks of passenger pigeons settled on branches and in the leaf litter to feed on chestnut mast. There was no better food for having the hogs put on sweet bacon. Chestnuts were stored away in salt to fend off insects and were eaten whole all winter long by Appalachian families, as by the Cherokees before them; pounded into meal or diced, they formed the basis for many a recipe, from breads to soups to stuffed turkeys. For the mountain people of Appalachia, chestnuts were practically manna from heaven, a great outpouring of food delivered at the doorstep at no cost. They were consumed locally and gathered for shipment elsewhere. Like moonshine, chestnuts could be readily sold or bartered; at the tail end of the nineteenth century a gallon of chestnuts equated to three to four pounds of sugar at general stores in Kentucky. From there they traveled up the socioeconomic ladder. In 1911 a single railroad station in West Virginia shipped 155,000 pounds of chestnuts to the cities of the East. Few other foods were as democratic as chestnuts: they were craved by people rich and poor, rural and urban. Roasted nuts smelled like butter wrapped in leaf litter and gave a satisfying crunch. In 1856 Thoreau visited New York City and found chestnuts "roasting in the street, roasting and popping on the steps of banks and exchanges. . . . The citizens made as much of the nuts of the wild-wood as the squirrels. Not only the country boys, all New York goes a-nutting."

Charlotte Ross, a folklorist at Appalachian State University in Boone, North Carolina, who is a doyenne of Appalachian studies, summed up the importance of the species to me this way. "It was everything," she said. "It was our economy, our culture. It was the reason the Appalachian region was settled, because we were a herding people and the animals fed on the chestnut mast. It's what we built our homes from. It was a large part of our foodways and of our folkways. Chestnut defined the region. If ever a region was associated with a tree, then the chestnut was *our* tree. People would hang on waiting to die not to get to another birthday or Christmas, but in order to go chestnut gathering one more time."

৯৽

WHEN THE CHESTNUTS DIED, the passenger pigeons hadn't been gone very long. Elk had been killed off east of the Mississippi and bison had been slaughtered in apocalyptic numbers out on the Great Plains, and you'd think people could have grasped the notion of a fantastically abundant species disappearing before their eyes. But, of course, no one had the faintest idea that such a thing could happen to the commonest of trees. Chestnuts were grocery and lumber store and shelter from summer thunderstorms all wrapped into one. Of course they'd always be around. But they weren't.

The end came with shocking suddenness. In 1904 a man named Herman Merkel discovered odd sunken cankers on dying chestnut trees at the Bronx Zoo. He showed them to W. A. Murrill of the New York Botanical Garden, who diagnosed the malady: a new fungus. That was good to know, but in the meantime the disease spread; by 1907 it had shown up as far north as Poughkeepsie, New York, as far south as Trenton, New Jersey, and east into Connecticut. Afflicted trees developed huge cankers and inevitably died within a year or two.

A few plant pathologists quickly recognized the magnitude of the problem and issued control guidelines. Chestnut trees and saplings were to be carefully inspected, eliminated from nurseries, and quarantined if necessary. A blight-free zone was to be established around the infected area. Arborists tried cutting off dead branches and carrying out surgery to remove cankers, but that was akin to shouting at the ocean to dissuade the surf.

The fungus eventually came to be known as *Cryphonectria parasitica*, and its life cycle wasn't hard to tease apart. Researchers learned that it killed trees by spreading fine threads through the sapwood and that it dispersed by means of fantastically abundant spores that were produced both in summer and winter. Summer's spores were sticky and clung to the feet of such birds as woodpeckers, which could spread millions at a time. Winter spores were motes of dust; they were emitted like those of a puffball and traveled on the wind.

The scrape of a bird's foot or the chewing of a tiny insect could open up the bark enough for scores of spores to enter. By 1911 the blight had spread to Massachusetts, Delaware, Virginia, and West Virginia, and officials in Pennsylvania granted the munificent sum of $275,000 to a newly appointed chestnut commission with the power to access any private land to look for signs of disease. It employed two hundred inspectors to look for signs of infestation. Its members considered ordering the clearing of a mile-wide swath—a sort of arboreal Maginot Line—in hopes of stopping the fungus. Most of the inspectors were trained in entomology, not fungal diseases, however, and it proved difficult to force landowners to remove diseased trees swiftly; anyway, the fungal spores traveled far too readily on wind or bird or boot. The commission's paltry defenses were overwhelmed.

In 1913 a federal plant explorer found the same fungus on a trip to China, where it grew without causing much harm on twigs of the native Chinese chestnut. That made it clear where the blight had originated, and the popularity in the nursery trade of Chinese and Japanese chestnuts—which produce nuts prolifically and whose low, stocky stature orchard growers liked—made it all too clear how it had traveled over the ocean. This knowledge, though, didn't contribute to controlling the blight either. By 1915 it had arrived in North Carolina and eastern Tennessee. In Virginia's Shenandoah National Park there were so many dead trees that bulldozers knocked them over to reduce the threat of fire. Some innovative park employees crafted drinking fountains out of the remnant chunks of wood. By 1926, when the federal government authorized the purchase of lands for Great Smoky Mountains National Park, lumbermen used the blight's presence in those rugged mountains as ammunition in an effort to defeat the park proposal. They lost, but so did the park's forests. In 1935 a few healthy stands of chestnut still grew there, but by 1938 they were all dying. All throughout the eastern mountains the bleaching trunks of dead chestnuts stood or lay like the bones of great vanished animals. Boys made a game of leaping from log to log, one to another, without having to step on the ground.

An estimated four billion trees died. It was an ecological and economic disaster and, Charlotte Ross points out, a deeply felt social and cultural one. "Old people look at you and say, 'It was as bad as the world wars and the Depression,' and they'll mean it," she said. "It was the loss of a subsistence lifestyle. 'You never saw the chestnut,' they'll say; 'sit down and I'll tell you.' And they'll say it with tears in their eyes."

Of course the mountain people did learn to adapt to their new ecological reality. Lumbermen quickly realized that chestnuts resist decay sufficiently that they could still be profitably cut as much as thirty years after dying. If boring insects attacked the wood, so much the better, as their burrows allowed chestnut lumber a better hold on the glue used in holding veneer together, and the so-called wormy chestnut, with its built-in guarantee of a limited supply, came to have its own cachet as a luxury wood.

The business that profited the most from the vast acreage of dead trees, however, was the tanning industry. Dry chestnut wood contained 11 percent tannin, and throughout Appalachia men fanned out into the woods to collect it. At the Champion Fibre mill in Canton, North Carolina, liquid tannin was extracted from twelve cords of wood per hour, twenty-four hours a day, seven days a week. The used pulp was mixed with pine and spruce to make paper. When the GIs stormed ashore on D-day the leather in their boots had been tanned with Appalachian chestnut, and their maps of Normandy were printed on paper imbued with chestnut fibers. If anyone perceived this as the New World taking revenge on the Old for the blight, he took cold comfort in the fact.

꿔

I SUSPECTED THAT BOB ZAHNER could tell me what it had been like to see the chestnuts die. He is a retired forestry professor who lives at four thousand feet just outside the summer resort town of Highlands, North Carolina. I got to know him and his wife, Glenda, some years ago in Tucson, Arizona, where they spend each winter. I found him on his home ground, wiry and white-haired but with a good deal of life in his blue-green eyes, on a lawn surrounded by tall hemlocks and yellow poplars and inset with a huge vegetable garden. He was in his eighties, but spry. Zahner's family began spending summers here in the 1920s, when his father was a successful Atlanta lawyer; they bought sixty acres of old fields and pastures atop Billy Cabin Ridge and converted an existing farmhouse into something of a mansion, some of it built of chestnut milled on site. Zahner's mother put in an informal rhododendron garden on the ridgetop; the old hillside pastures filled themselves in with trees. Zahner and his brother Kenny used to walk up nearby Whiteside Mountain in autumn and filled up their pockets with chestnut burrs, which they called "porcupine eggs." They roasted the nuts at home. Then World War II came and Zahner became one of those GIs roaming around Europe in boots tanned with chestnut extract. For three long years he carried the memory of the woods with him. After he came back he built his own house just down the hill from his parents', a rustic two-bedroom cabin, gathered around a central fireplace, that came to be paneled as much in books as in dark wood.

Now he was retired and needed to carry an oxygen pack sometimes to get around; he felt the elevation, and although Highlands appears to have a pristine mountain climate, a good deal of air pollution drifts there from all the Tennessee Valley Authority power plants to the west. "It's not fair," Zahner said, "I never smoked a day in my life." His parents had long since passed away, the big house had burned down, and some of the land had been sold off by the family, but Zahner's son lived in a neighboring house now, and the hand-built cabin was much the same. He and Glenda still live there when they're not in Tucson. It was on one of its two porches that we drank coffee and talked about chestnuts. Zahner remembered the last one that grew here very well; in fact, he could show it to me. It was just down the hill, below the hemlocks, in what had been pasture when the Zahners had arrived.

"There were a few big chestnuts and oaks that had been left in the pasture for shade," he said. "Chestnuts did well on these comparatively dry sites. They didn't form gigantic trees around here, but they did quite well. The last one died about 1938. It was a beautiful snag and only fell over a few years ago." When we finished our coffee, we walked the fifty yards down the hill to where the great tree lay—a shattered giant among the trunks of much smaller hemlocks, tulip poplars, and rhododendrons. The great bole, silvered on the outside but turning punky and reddish wherever the rain was able to enter, was more than four feet in diameter. It had broken off a few feet above the surface. The log ran straight for twenty feet before the first branches emerged. They were themselves a foot in diameter. The trees were tall all around us. It was difficult to believe that the place had ever been an open meadow where hardscrabble farmers watched over lowing cattle.

"It turned out that the chestnut blight coincided with the Great Depression of the 1930s, and cutting the dead chestnut was a boon to the local people," Zahner was saying. "A man and his brother could take a mule or a horse and get a permit to harvest dead chestnut. With a horse or a mule they could go up into the most inaccessible places. They'd make sleds out of sourwood trees and put a team of horses on them and haul them up into the most inaccessible woods. They'd cut the dead chestnut into six-foot lengths and split them with their axes and wedges into manageable sizes. Then they'd haul them on their sleds to some place where they could get an A-model truck in. All these county seats had plants where they would extract tannin from the acid wood, and the men could sell the wood to them. It was an economic salvation for the poor people up here in the hills. Thousands and thousands of families made their living that way up here. That's the only good thing that came out of the loss of the chestnut. Every one of the mountaineers I talked to around here depended on acid wood. That's why you don't see many snags any more, but lots of stumps."

He had his own, more personal connection to the dead trees. In 1950, after all the chestnuts in the area had died, he and Kenny cut a moderately sized one from their land and hauled the biggest section of trunk they could handle to the Highlands Biological Station, a nature center in town. It's still there today, a robust snag, frozen in time, that barely fits inside the building. We went to look at it that afternoon. It's a lot smaller than the one that lay in the woods.

What he really wanted to show me, though, were the ancient hemlocks that grew in a wooded cove out behind the building, on a tract of land that had never been logged. They towered overhead, filtering the sunlight: trees a hundred feet and more high, a good four feet in diameter. They had a complicated architecture of large branches and ragged tops and, in a few cases, trunks broken as a result of wind or a lightning strike. They grew high above thick-trunked laurels and magnolias: sylvan giants, a bit of Appalachia's original old-growth forest left standing in a sunken grove hidden among the summer houses and golf courses. Warblers sang buzzy songs in their shade.

The hemlocks now were all dying. Showing them to me was a bit of time travel on Zahner's part. In recent decades the hemlocks of eastern North America have been suffering from an imported pest every bit as virulent as the chestnut blight. Looking at these trees was like standing in a grove of towering chestnuts in 1935. It was the end of an era.

The pest was a tiny insect called the hemlock woolly adelgid, which like the chestnut blight fungus had inadvertently been imported from East Asia on nursery stock. It had been working its way up and down the eastern mountains since the 1960s and got to Highlands about 2000. Adelgids live by sucking the life out of individual hemlock needles, and Zahner showed me how to spot them by looking for the white fuzz they secrete around themselves. It is possible to release a beetle that feeds on the adelgids, he said, but its use is complicated and is likely to be successful only in small tracts.

The bases of the lower needles of the hemlocks in the cove were all covered in white fuzz. When the wind and the warblers and woodpeckers grew silent for a moment, I could hear the hushed sound of needles falling. Within a few years, Zahner said, all these hemlocks would die. He didn't know what would happen to the warblers that preferred nesting in them: would they be able to use other trees, or would they simply disappear from the area?

These great living trees conveyed much better than the fallen chestnut snag just what had been lost back in the 1930s. A dead tree is just a dead tree, but a living tree soon to die is an elegy, full of life and death alike. The hemlocks were, I thought, a sad look at a possible future of ecological restoration: in a world of global trade, where fungi and insects and invasive plants can be carried halfway around the globe in a matter of hours, new die-offs like that of the chestnut are all but inevitable. So, perhaps, are the efforts of people to fix what's been broken, to return the land to health.

"I don't have a clue what you can do as far as restoration of the hemlock goes," Zahner reflected, "but then no one knew that about the chestnut years ago either."

ॐ

WHAT FORESTERS *did* KNOW ABOUT CHESTNUTS, back in the day of the great dying, was that Chinese and Japanese chestnuts were much more resistant to the blight fungus than the American species. They had evolved with it, and although they were commonly afflicted they generally did not die. Armed with that slim fact, plant breeders around 1930 began crossing blight-resistant Asian trees with some of the few surviving large specimens of American chestnuts. If a hybrid tree survived, it was crossed again with another Asian tree. No one knew precisely which parts of the Asian trees' genetic heritage conferred resistance from the blight; no one knew much about genetics, period. Not much remained of the genetic heritage of the American trees. Asian chestnuts are much shorter than American trees—their size and stature are closer to those of apples than of oaks. Not surprisingly, the hybrids were short and stubby, and most didn't survive the blight anyway.

By the time World War II was over, interest in chestnuts was itself dying off. The tannin industry took care of most of the dead trees, and the people of Appalachia came to focus on economic development and the new postwar consumer economy. The subsistence-centered lifestyle once represented by chestnut trees quickly came to seem part of a distant, impoverished, and not-so-desirable past. There were dams and coal-fired power plants to build, cars and washing machines and televisions to buy. The places where chestnuts had grown filled in with oaks and hickories, maples and black locusts; the forest canopy closed in, and it all seemed as good as new to young people who hadn't witnessed the blight. Every year memories of the big trees and the big harvests grew fainter, borne back ceaselessly into the past like grainy black-and-white recollections of bad roads and lean winters and the hardscrabble but resilient economy of the old days. Efforts by the United States Department of Agriculture (USDA) to breed a blight-resistant tree were discontinued about 1960, while a breeding program at the Connecticut Agricultural Experiment Station continued to plod along with few overt signs of success.

Curiously, the species itself survived. Before the blight struck people had spread chestnut seeds widely, and fungus spores were not able to reach all these places. Substantial groves thrived through the twentieth century in such places as Michigan and Wisconsin that were outside the native range of the chestnut (today the largest tree of the species grows in Oregon). Even within the original range a few large trees persisted, either because they happened to escape infection or carried some degree of resistance.

And throughout the woods new sprouts sprang up. Chestnut roots seem practically immortal. As Thoreau indicated, when a trunk dies the extensive root system is liable to send up an array of new stems. "It keeps smoldering at the roots/And sending up new shoots," is how Robert Frost put it. The blight fungus attacks only trunks and branches, not roots, so wherever it had passed through, new shoots sprang up, only to be attacked themselves when their trunks reached a diameter of several inches.

That characteristic has led to a paradox: within their old range, chestnuts are functionally extinct as a large tree and a source of nuts, but, in fact, are quite common. When an intern with the American Chestnut Foundation through-hiked all 2,015 miles of the Appalachian Trail a few years ago, he tallied some thirty-five thousand chestnut sprouts along the way. I saw some there myself near Newfound Gap in Great Smoky Mountains National Park, not far from where a couple of hikers asked me, "Chestnuts? Aren't they extinct?" Because its spores are spread largely by wind, *Cryphonectria* does not disperse well in dense forest, and so throughout much of their old range chestnuts exist in a sort of smoldering vegetative quiescence, safe from the blight as long as they remain in the shade and grow slowly and not too large. They're not dying out, but they're not reproducing either, so they remain unable to explore new genetic possibilities through the crapshoot of evolution. They have a chance to bloom only if hit with an infusion of light—which was why the chestnuts I saw that rainy afternoon in the clear-cut near Meadowview had responded so vigorously. Then they, like virtually all other chestnuts, were hit with the blight within a couple of years, and only through Hebard's intervention were they able to hang on long enough to pass on their genes.

In 1981 a geneticist named Charles Burnham, who was in his seventies and on the cusp of retiring from the University of Minnesota, wrote a letter to the professional journal *Plant Disease* suggesting that the USDA's chestnut-breeding program, begun some fifty years earlier, had been misbegotten. The strategy was wrong, he wrote. Instead of crossing Asian-American chestnuts with more Asian trees, he said, they should be backcrossed with American trees, and then those that survived should be backcrossed again. That way the Asian genes for disease resistance would be passed along, but the remainder of the trees' genetic heritage would come to more and more closely resemble that of purebred American trees.

This technique for speeding up the glacial process of evolution had been proven to work with wheat, corn, and barley. It wasn't easy, and it wasn't quick—it had taken agricultural researchers fifty years to develop a variety of wheat that was resistant to black stem rust—but it proceeded with the inexorability of a mathematical equation. Burnham was fairly sure that a chestnut tree's resistance to blight was governed by only two or three genes. That allowed him to calculate precisely how many hybrid trees would be needed before probability dictated that some would express, in their random genetic shuffling, the Chinese genes needed for resistance amidst an otherwise American heritage. Unlike corn, which after endless breeding was enormously productive but also required fertilizers and tractors and pesticides and the careful ministrations of legions of farmers, he intended to produce a reengineered plant that could move into the woods on its own and become self-sustaining once again.

Burnham was a colossal figure in the world of plant genetics. Although he'd made his mark by studying the breeding of corn, he had borne an interest in chestnuts since the 1930s, when as a freshly minted Ph.D. he saw remnant chestnuts bearing flowers at a postdoctoral posting at the University of West Virginia. His letter came as a shock to all the plant geneticists who had been taught in school that the case of the chestnut proved that breeding for disease resistance did not work.

"This was a scientific problem that the entire world scientific community had announced was finished," Philip Rutter told me. He has been breeding chestnut hybrids on his Badgersett Farm in southeast Minnesota since the 1970s. "We couldn't do anything about it," he said. "It was considered an absolutely lost cause, and academic suicide to work on it. This was a dead horse if ever there was one."

The letter didn't shock Rutter, however. He was a graduate-school refugee who was trying to figure out, with his wife, how to make a living on the farm they'd bought as a cheap investment. Growing chestnuts for food seemed one possibility. This mutual interest brought him to Burnham, and the two men became close friends. They spent two years discussing the mathematics of breeding, and Rutter came to agree that Burnham's proposed backcross method was feasible. The moment of truth came when the two men were sitting in the kitchen of Burnham's house, eating grilled-cheese sandwiches.

"We looked up at each other and said, 'This can be done,'" Rutter said. "So then the question was, What do we do?" Burnham broached the possibility that the USDA or the U.S. Forest Service might begin a new breeding program. No, said Rutter, what was needed was an organization solely focused on chestnuts, "one that could live longer than either one of us."

Having talked the solution up for two years, Rutter could not help but become president of the new American Chestnut Foundation (ACF) in 1983. Burnham became the new organization's eminence grise and lent it scientific credibility. Rutter learned how to raise funds and run a nonprofit organization. He also oversaw some early plantings of hybrid chestnuts at several universities, but it was too hard to coordinate them. What was needed, he thought, was a farm that also could live longer than any individual.

In the late 1980s he got lucky. Rutter happened to give a talk about chestnuts to an audience in Maryland that included a woman named Anna Belle Wagner and her two daughters. They owned a farm in western Virginia that they didn't really know how to use. They offered it to the ACF. The foundation took over a long-term lease, with an option to eventually buy, from the Wagners. By the time Charles Burnham died in 1995, the breeding program he'd dreamed up was well under way at Meadowview's Wagner Farm. Promising results were starting to show.

⁂

FRED HEBARD GREW UP IN SUBURBAN PHILADELPHIA, not far from farms and fields. He craved the rural life. He went to college for a while but dropped out and began working on a dairy farm in Colchester, Connecticut. One day he accompanied the farmer he worked for in chasing a heifer that had strayed into the woods. "We came across a chestnut tree—a sprout—and I thought it would be a good idea to go back to college and cure it," he said. "I didn't know it was a lifetime proposition."

Hebard was twenty-two. He went back to college and graduated from Columbia University in the biological sciences. Next came a master's in botany at the University of Michigan, and then a Ph.D. at Virginia Tech in 1982. His advisor was a plant pathologist named Gary Griffin, who guided his student's studies on the histopathology of chestnut blight. Next came a postdoc at the University of Kentucky, which also involved work on the disease's physiology. In 1988, when he heard that the ACF was looking for someone to work on the breeding of disease-resistant chestnuts at a newly acquired farm in Meadowview, he jumped at the chance.

It wasn't an easy choice. The pay was poor, and Meadowview was a long way from anywhere. Hebard's wife had also studied plant pathology, and there was no job available for her. They and their two baby daughters would be living in an old farmhouse with slanting wooden floors that doubled as an office for the ACF. It would, though, be a chance to guide the restoration of the chestnut and, for Hebard, an opportunity to do hands-on work in what he called a "practical, field setting." He'd still spend some of his time peering through a microscope, but he'd also be out with the trees, in all weather. He took the job.

"The problem is in biology there's no distinction between scientist and engineer," he told me over lunch at the Little Diner. "I consider myself basically an engineer, but there's no term for that." Whatever the term, Hebard has, since his arrival in Meadowview, been consumed with both the theory and practice of breeding chestnuts.

When I went to talk to him on the morning before it rained he was walking on the Glenn C. Price Research Farm through an orchard of five-year-old chestnut trees, most between twelve and twenty feet high. They were third-generation backcross trees. They were the result of a series of controlled pollinations. First, a Chinese chestnut tree was crossed with an American tree. In this case the product of this breeding was the "Clapper" hybrid, which had been developed during the old USDA breeding program. Hebard had then chosen trees of the next two succeeding generations that displayed good resistance to the blight and had crossed them with pure American chestnuts. In theory this scheme meant that the end result was trees with a genetic heritage fifteen-sixteenths American and one-sixteenth Chinese.

That was the theory; in fact, any parent knows that genes are shuffled a bit more wildly than that during mating. The fractions are only averages. What Hebard was doing that morning was culling the third-generation trees that didn't pass muster, leaving to stand only those that had both good blight resistance and a predominance of American chestnut characteristics.

The first choice was easy. Trees that weren't resistant to the blight were simply dying. Fred pointed one out. Its trunk was about five inches in diameter, and about a foot off the ground a canker that looked big and ugly and sunken to me—it was the size of a saucer—had spread around it. Above that the trunk split. One of the halves of the crown was already dead. The leaves on the other half were a sallow yellow-green and were small. Hebard used a pocketknife to cut away the bark at the canker's edge. Outside the canker the cambium—the living wood just below the bark—was white and juicy, while inside it was brown, dry, and dead.

"You see how the canker extends all the way around to the side?" he asked. "We won't breed this tree because it doesn't have enough resistance. So we'll take this one out with a backhoe."

Genetics is all about chance, and because he wants it to proceed expeditiously Hebard tries to leave as little as possible of the rest of the chestnuts' lives to chance. He weeds and he fertilizes, and he doesn't wait around until trees are infected by fungus by a chance puff of wind; he deliberately infects them. This tree had been injected with killing fungus two years earlier. I'd seen how it was done the previous day when I helped out the Elderhostel group as they moved through an adjacent grove of four-year-old trees. They punched two small holes as big around as a pencil eraser at the base of each tree and filled each hole with a clear, gelatinous mass of fungal culture from a petri dish. Then they wrapped the wounds with white first-aid tape. The two holes were filled with *Cryphonectria* strains with different degrees of virulence. When Hebard returned in a year or two he would measure the resulting cankers and assess the condition of each tree to evaluate which had the best chance of passing on the genes for resistance.

Hebard's second decision—choosing the trees that had the greatest proportion of American chestnut traits—was trickier, for it involved assessing an entire palette of characteristics. He was looking for leaves that were shaped like canoes, whereas Chinese leaves were broader; that were rather dull, rather than the high gloss of the Chinese; that weren't hairy on the underside; that had long teeth "like a curling ocean wave," as he said; and whose veins had only a few hairs. He was looking for trees whose bark had lots of small lenticels—tiny holes through which a tree absorbs carbon dioxide—rather than a few large ones. He was looking for trees that rose high, rather than compact and stubby ones.

A tree with all those characteristics would be allowed to mate. That too was a complicated process. Down on the south end of the farm I could see large chestnuts whose crowns were dotted with white cotton bags. The bags had been draped over chestnut flowers so that uninvited pollen grains would not fertilize the female flowers; Hebard called them "chastity belts for chestnut trees." Instead, Hebard and his crew would dust those flowers with pollen gathered from selected trees in the next few weeks.

Those trees that Hebard didn't select were left for the backhoe, which was already grinding away at the other end of the field. The trees it uprooted were dragged off and burned. "People ask if they can get the stuff for lumber," he said, "but I've declined—it would be just another chore."

That afternoon, once the backhoe driver had done his work, there would be only a few scattered trees left standing in what had just a year or two earlier been a fully stocked orchard. They were, in theory, quite Americanized and quite blight resistant. They were important breeding stock, but they were not the end goal of the breeding program.

They represented, for one thing, only a single genetic heritage, and Hebard wanted to breed trees that would express the full range of the chestnut's genetic variability. He wanted trees that could deal with the heat of Alabama and the cold of Maine, with moist coves and dry ridges. He wanted trees that were tall and short, gnarled and straight-trunked, copious or stingy with their nuts. He wanted chestnuts, "warts and all," as he put it, because if the trees he bred were too closely related, too inbred, *Cryphonectria* might quickly do its own evolving and find a way to kill off the new hybrids. For that reason, he was trying to breed numerous hybrids with varied lineages. A key to which trees are planted where in the more than fifty separate groves on the Price Research Farm takes up an entire page and consists of cryptic notations such as "Maines 2001. Clapper third backcrosses 3/9/01" and "Pauls 2002. Various F2s 3/15/02."

It is because of this desire to maximize genetic diversity that the ACF has chapters in many of the states in the chestnut's old range. In all those states, from Maine to South Carolina, volunteers have been conducting their own breeding programs—often with pollen shipped overnight from Meadowview during the busy days of June—to craft hybrids suited to those sites. "The best thing about this whole project is the volunteer spirit," Paul Sisco, staff geneticist for the ACF, told me. "There are people who have chestnut signs plastered all over their trucks. There's a real energy. I know of one woman who devotes all her vacation time every year—all of it!—to pollinating chestnuts."

There was also no guarantee that the third-backcross trees would breed true, because they had been getting a new infusion of American genes with every cross. Their progeny might inherit either Chinese genes for blight resistance or American genes for blight susceptibility. To ensure that the genes for resistance were passed on, the resistant hybrids had to be intercrossed with one another for a further three generations. Genetics works in an exponential manner. Of the third-backcross trees Hebard was assessing, some 12.5 percent were somewhat resistant and 87.5 percent had low resistance. But after those hybrids were bred with one another for another two generations, says the theory, 1.6 percent of the resulting trees would be fully resistant and 64 percent somewhat resistant. After one more generation, 100 percent would carry the same degree of disease resistance as their Chinese ancestors—that is to say, most would survive. And because the American genes for blight susceptibility will have been bred out by then, those trees should breed true.

Hebard had, in 2002, planted the trees that would bear those 100-percent-resistant nuts. They were on an eighteen-acre tract, newly acquired by the ACF, a few miles from the Price Farm. It didn't yet have an official name because no donor had yet stepped forward to underwrite its purchase, but the local farmworkers called it the Duncan Place after a previous set of owners. We went there just before heading up into the Jefferson National Forest in the rain.

With its many midsized trees, the Price Farm looked like an orchard. The Duncan Place looked like science. Chestnut seedlings one and two and three feet high poked up from tightly arrayed aluminum tubes, row upon row upon row. Pokeweed and thistles and poison ivy sprouted from some of the tubes with the young trees, and here and there Hebard stopped to pull out a weed stalk. The key to making chestnuts flower, he said, was "plenty of water, fertilizer, lots of sunshine, and no weeds." He and his crew would be here frequently in coming years both to combat the weeds and to cull the chestnuts. These thousands of seedlings would be thinned down into only about thirty trees per acre, which, if Hebard chose well, would represent the 1.6 percent of the population with full disease resistance.

It is the progeny of the trees at the Duncan Place, which will probably be produced in good numbers beginning in 2006, that will represent the acid test of the ACF's more than twenty years of chestnut breeding. The ACF plans to distribute them to volunteers throughout chestnut country, who will plant them at selected sites in the woods. As every farmer knows, it's one thing to breed a plant that can thrive in a field; it's a very different matter to breed one that needs no care at all.

His goal, Hebard said, is "to get enough genetic diversity so they can get out there and reproduce on their own. If they have too much Chinese they just won't be able to compete. And I plan to be here to see it."

That may have been hyperbole, but Hebard does, in fact, stand a good chance of witnessing at least the initial success or failure of his breeding program. Because chestnuts grow so quickly, he's in an unusual position for a tree breeder. The usual pattern, Paul Sisco told me, is that "you do your breeding, establish your seed orchard, and then retire." Hebard is young enough that he should easily be able to assess how the progeny of the Duncan Place trees do—if not when they're three feet in diameter, then at least at six or eight inches. It's a nerve-wracking prospect to wonder how the mathematical precision of genetics will bear up when faced with the uncertainties of climate, insects, disease, competition from other plants, and all the other variables of life in the woods. "Fred's getting a bit nervous about whether the trees are going to be resistant enough," Sisco said. That was in 2004. "That's where the rubber hits the road. It's always easy to say it'll be another five years down the road."

₰

THE MORE IMMEDIATE QUESTION, though, may be not whether the ACF's hybrids will do well in the woods, but whether it is a good idea to put them there at all. They will be hybrids, after all, with a genetic legacy a little bit unlike the old American chestnuts. That bothers some people. One of them is Hugh Irwin, a forest conservationist who works for the Southern Appalachian Forest Coalition in Asheville and also serves as the chair of the ACF's science cabinet. In 2004 he told me he was worried that the Duncan Place progeny would not be American enough to truly meet the goal of restoring the native species, and he called for additional generations of backcrossing to increase the percentage of American genes.

"I don't feel that the third backcross will necessarily be a good candidate for getting back into the wild," he said, "though it will be a milestone, and very effective for further testing. I have advocated for a more measured approach—more testing, and moving the backcross component forward so that we approach something like 99 percent. I feel that we have time to do it right, to get it to a place where it pretty uncontestably will be chestnut restoration, as opposed to a hybrid that's *mostly* American."

At the core of Irwin's critique was concern not that the hybrids won't thrive, but that they will thrive too well while not closely enough replicating what they are intended to replace. Hebard believes that hybrids with too large a component of Chinese characteristics will simply not survive in the woods; they would, for one thing, not grow tall enough to compete with other species. It is the American genes, he maintains, that will make chestnuts able to grow in the forests of the East.

Irwin is intimately familiar with the influx of invasive plants into eastern forests, from kudzu to tree of heaven to Japanese honeysuckle, and he worried that the new hybrid could act as aggressively in competing with native species as those introduced varieties have. "There could be a hybrid that exhibits American characteristics, sort of, and is blight resistant," he said. "When we get it out in the woods we realize it's kind of American but exhibits other characteristics too, and it would be too late to stop it." Such hybrids, he worried, might swamp the native American genetic potential of all the chestnut sprouts that continue to smolder—albeit in an unreproductive quiescence—out in the woods.

It is, indeed, the sheer vigor of chestnuts that lends credence to Irwin's concerns. If they can get around the blight problem somehow, chestnuts grow fast. "We're trying to restore a real aggressive species," said Paul Sisco. "Just introducing it is like introducing a time bomb." He referred to a chestnut stand near West Salem in southwestern Wisconsin where a farmer named Martin Hicks planted nine chestnut trees along his fence line in the early years of the twentieth century. Because the trees were well outside the chestnut's natural range, they remained blight free, and by 1988 they had proliferated into thousands of trees that covered nearly sixty acres (that year, *Cryphonectria* somehow made it to West Salem, and the trees have been in decline ever since). At a blight-free site in southwest Wisconsin, test plantings of American chestnuts in the mid-1990s showed that those trees readily outcompeted native black walnuts and red oaks by growing almost a yard in height per year.

Advocates contend that getting the hybrids off the plantation and into nature's hands is exactly what is needed. Phil Rutter believes that the breeding program is only a variation of what nature would do on its own, given enough time. "The plain fact of the matter is if the American chestnut had been able to recover from the blight on its own, those trees would not be identical to the original either," he said. "This is natural evolution. The species has come through a genetic bottleneck, and any species that does that will be slightly changed. But it *is* an American chestnut.

"Life moves. Species change. This is a matter of us fixing a chestnut problem with chestnut genes that happen to come from another species. It's no different than natural evolution except that it's a bit faster. The fact is that *Homo sapiens* is the major evolutionary force on the planet at this point."

At its core, the argument is about purity, and it is a conflict that dogs numerous restoration efforts: How closely should an ecosystem —or an organism—that people re-create hew to the original? How cautious ought people be in shaping nature? The argument persists because it is impossible to resolve. Any measurement of purity, after all, is arbitrary. Is it sufficient to carry 93.75 percent of the genetic heritage of the original? How about 99 percent? Is the measurement of gene percentages even a useful measure of the extent to which a hybrid can replicate the ecological role of the original? Answers to such questions don't have much to do with the precise equations that govern genetics; they're based more on ideology and values than on science.

"The folks who are what I would call purists, there's just no satisfying them," Paul Sisco told me. "For some of us that grew up in the South [he grew up in Memphis] it's reminiscent of the old belief that 1 percent of Negro blood made you a Negro. It's the same kind of talk about purity. It's almost a religious belief, and nothing you can shake."

Irwin didn't argue that these questions should shut down the chestnut restoration project, merely that the work should be done as carefully as possible and that potential consequences should be thought through beforehand. He was simply articulating a higher standard of caution than did Hebard and his colleagues. When Hebard, the pragmatic biological engineer, looks over the straight rows of chestnuts planted at Meadowview he sees hybrids with the potential to bring a source of nuts back to the eastern forests and double the wood production in Appalachian forests, and he envisions nature further sculpting their genetics, generation by generation, into individuals increasingly tailored to the specific local conditions of their habitat. When Irwin looked at the same trees he saw products of farming that should not, as yet, be mixed with wild places because their genetic heritage isn't quite consistent with the ways in which wild plants evolve. He thought that people should sculpt the trees' genetics a bit more carefully before setting them loose.

By early 2006 Irwin said his fears about these particular chestnuts had been allayed, and what was likely to happen was that chestnut hybrids will be planted on private lands and some public lands, while managers of the most ecologically valuable preserves will withhold judgment for a while. Greg Eckert, a restoration ecologist with the National Park Service, suggested to me that protected areas such as Great Smoky Mountains National Park might serve as reservoirs of the chestnut's native genetic diversity rather than as experimental homes for the new hybrids. "Maybe we should not pollute the gene pool in the national parks," he said. "I'm not sure we should do that. Lots of people will be planting hybrids, and maybe we should play a different role. There's value to preserving the species' original genetic makeup.

Eric Higgs, an environmental philosopher at the University of Victoria in British Columbia and for a time president of the Society for Ecological Restoration International, has seen many restorationists wrestle with these issues. "For example, there's a local stream restoration project," he said, "and, boy, is it ever hard to get any good historical information. It's really tough. It was a small site, and it was pretty unglamorous. No one took many photographs of it. You comb the archives, you go back in time, talk to neighbors—not much, right? So what do you do with that? I think the process of trying to uncover the history of that site helps increase the sense of responsibility you have to be careful in what you do—to recognize that your intervention is part of the historical sequence, and you're trying to do things differently.

"What restorationists are really good at is understanding ambiguity. Restorationists have got to revel in that kind of ambiguous terrain where things are neither this nor that."

ॐ

WHEN IT COMES TO CHESTNUTS, Lucille Griffin does not hold truck with ambiguity. In and around Blacksburg, Virginia, a small college town, she administers to hundreds of chestnut trees, and if most of them are quite small, they are all indubitably American. She is the director of the American Chestnut Cooperators' Foundation (ACCF), which has the goal of returning trees that are blight resistant and of purely American stock to eastern forests. To that end, she oversees a breeding program carried out by volunteers that spans multiple states. It's a mirror image of the work the ACF carries out, except without the addition of genes from Asian chestnut trees—and without a big budget. The ACCF operates on a shoestring and, to a considerable extent, on Lucille Griffin's shoe leather.

I visited Griffin one sparkling June morning, on a day so lovely that she said it reminded her of camping trips to Maine she and her husband used to take with their children. Lucille's husband is Gary Griffin, the Virginia Tech chestnut researcher, semiretired now, who was Fred Hebard's graduate advisor. Some two decades ago the couple's four children were almost grown and Lucille didn't have as much reason to go camping or to grow a huge vegetable garden, and one fall she saw that an experimental plantation of chestnut trees that Gary had planted in Blacksburg was dripping with nuts and there was no one to pick them. She thought that was a waste and picked them herself with her father, who was visiting.

Griffin wrote an account of her experience for the *Wall Street Journal*, which to her surprise published the piece. To her further surprise, she subsequently received a good deal of mail from readers interested in chestnuts—enough to initiate a network of cooperators who agreed to attempt the propagation of blight-resistant varieties. They would use only American trees as breeding stock.

"We do not deal with hybrids at all," was practically the first thing Griffin told me that day. She is a slender woman with silvering hair and a residue in her voice of Long Island, where she grew up. "Our breeding program from the start has been breeding for blight resistance in American chestnuts. We call them all-Americans for short."

To show what the results of that work were, she took me to a clearing on the Jefferson National Forest that had been clear-cut a few years earlier. A Forest Service staffer had alerted Griffin to the quantities of chestnuts sprouting there, and now, under her care, the clearing was dotted with chestnut trees eight and ten feet tall. The trees were surrounded by round wire cages intended to exclude rabbits and deer. Voles, which avidly devour sprouting chestnuts, were harder to keep out. White strips of first-aid tape encircled many of the thin stems, giving the clearing the air of an intensive care ward for trees.

It was. The personal care Griffin administered here was exactly what distinguished her procedure from Hebard's. Rather than leaving her chestnuts to the vagaries of the genetic lottery, she tries to ensure that as many individuals as possible make it. She does so by using the same science that allows people to avoid deadly cases of smallpox or measles through vaccinations—by inoculating blight-ridden trees with a variant of *Cryphonectria* that doesn't kill them, at least not quickly.

"Hypovirulent" strains of *Cryphonectria parasitica*, as they're called, were discovered in Europe in the 1960s when an Italian researcher noticed that many European chestnut trees in that country were recovering on their own from the blight. Hypovirulent strains are of the same blight-causing species, but they lack the quick-killing effect of virulent ones. When they are injected into an infested tree they can take over the killing strains through a sort of cellular hijacking, allowing the tree to survive for a few extra years.

Researchers were excited to discover hypovirulent strains of *Cryphonectria* in the United States in 1976, but they simply haven't been as promising as in Europe (European chestnuts tend to be less susceptible to the fungus than American ones, so their recovery has been much easier). Still, in Griffin's careful hands inoculated trees can survive long enough to bloom and bear fruit within only a few years. Those that don't Griffin cuts back to the ground so that she can graft new woody shoots onto the roots. She doesn't like to waste the genetic and energetic potential of a good set of chestnut roots. "Their roots go way down," she told me. "You can't even imagine—they have a long, long taproot."

Griffin has been able to goad her wards into producing a great many chestnut seeds that she either plants herself or mails to her wide network of collaborators, some of whom do yeoman's work in planting, grafting, and inoculating in states throughout the chestnut's old range. Still, the work is painfully slow. Grafting demands finesse, and it is not uncommon for a grafted tree to break at the point of junction, months later, even if it had appeared to be in perfect health. We saw one such tree later that day—it was a twelve-footer, with its leaves still brilliant green, that lay on the ground, its trunk splintered. The tree was of a variety named for the son of one of her most reliable cooperators, in Tennessee. Even after twenty years of such experiences, Griffin was a bit taken aback.

"Oh, that's so discouraging," she said. "That's sad. And it had flowers on it. Oh, well. You can see how this really is two steps forward and one step back. You've got to be an optimist in this business. I've learned not to cry when my babies don't make it."

Griffin sees promise in this breeding approach even as she remains intimately aware of its painful slowness. "The breeding has the chance to get better each time," she said, "though it's just a matter of which traits combine, and you never know. Everyone said this was a waste of time. They said there was no blight resistance in the American chestnut. Now they recognize that there is just a lower level of resistance in it. Resistance in crop plants may take generations to express itself."

Whether those incremental degrees of resistance that might be accruing, generation by generation, in Griffin's breeding program will ever result in trees that can survive by themselves in the woods is very much an open question. "I think it's a reasonable program," Paul Sisco told me, "but I don't know how long it's going to take." Fred Hebard liked the ACCF approach when he began working with chestnuts, but then decided he wouldn't pursue it. He mused about it as we drove back toward Meadowview from the clear-cut he'd shown me in the rain. "When I first came I was more in favor of working with American trees than the crosses with Chinese," he said, "but as far as I can tell right now the backcrosses will work, while . . . " he trailed off and was silent for a little bit. "Now if you could get enough hypovirulence out there, that might work too." He didn't want to say anything negative about Griffin's approach, but he did want to see blight-resistant chestnuts growing out in the forest in his lifetime. He thought the hybrids he was creating at Meadowview had a much greater chance of exhibiting sufficient blight resistance, and sufficient genetic diversity, to survive in the wild.

⁊ᴼ

OCCASIONALLY, the Griffins get a call from someone—usually someone who's just learned about chestnuts—who reports finding a large chestnut tree thriving in a yard or woodlot. Usually it ends up being a Chinese chestnut. Lucille Griffin used to fantasize that one of these calls would result in pay dirt, in the discovery of an American chestnut tree that was entirely blight resistant and that could, by itself, resuscitate the species. "I never think that any more," she said. "I'm entirely disabused of that notion." If chestnuts were to survive as a mature forest tree, she now believed, it would be due to the hard, ongoing work of many people like her. Maybe Fred Hebard would succeed in getting his hybrid chestnuts to survive in the Appalachian wilds in coming years, and maybe they would acquire the tall stature and nut-producing vigor of the original species, or maybe not. In either case, she intended to continue grafting and inoculating—and hoping for signs of resurrection.

The Griffins live on an inholding in the Jefferson National Forest, a few miles from Blacksburg. Their property sits almost at the top of a long, uplifted ridge that looms over Poverty Creek like the crest of a breaking wave. From their house I wandered along the highway shoulder up to the top of the hill, from where I could see narrow and deeply forested ridge after ridge after ridge in the clear afternoon light. I tried to imagine how they'd looked when they were all dominated by chestnuts, not so long in the past, the exhalations of their leaves turning the far distances a hazy blue, or their June blossoms so thick that they turned the hillsides white as snow.

Griffin had told me how her interest in nature had awakened when she was a girl. Her family lived on the north shore of Long Island, one town over from Oyster Bay, where the families of men who owned banks in New York City lived. It was a beautiful, expensive pastoral—Great Gatsby country. Her father was a contractor who put in driveways and lawns for the wealthy. He loved the outdoors, and his daughter quickly became a tomboy. Often she played on a nearby farm where a stand of copper beeches grew. The beeches had stout branches that circled around their trunks like the steps of a spiral staircase, and an athletic child could climb to the very top. From up there she could see all around, even across the sound to the Connecticut coastline. The wind soughed in the branches and the distant shore was green and the sunlight and the white sails of yachts danced on the water. It was as close to heaven as one could get on Long Island. Later, the farm was bought as an estate by a wealthy family, and when the parents went yachting Lucille babysat the children and showed them her special stand of trees.

Standing there atop the ridge, I thought of the young Lucille climbing those grand beeches and falling in love with trees in the process. I was reminded of F. Scott Fitzgerald's great novel of longing and hopeless dreams. "You can't repeat the past," Nick tells Gatsby at one point, in the heat of the latter's quest for Daisy.

"'Can't repeat the past?' he cried incredulously. 'Why of course you can!'

"He looked around him wildly, as if the past were lurking here in the shadow of his house, just out of reach of his hand.

"'I'm going to fix everything just the way it was before,' he said, nodding determinedly. 'She'll see.'"

Of course, Daisy doesn't see, and Gatsby can't fix anything the way he wants it. In the end Nick returns to the mansion after Gatsby has died, and his dreams of regaining Daisy's love, never realistic, are as thoroughly gone as the curious and acquisitive light shining in the eyes of the Dutch sailors who'd first glimpsed that promising verdant shore, centuries before. Nick looks out at the sound and remembers "Gatsby's wonder when he first picked out the green light at the end of Daisy's dock. He had come a long way to this blue lawn and his dream must have seemed so close that he could hardly fail to grasp it." It was the dream of fulfilling the promise of the past, of achieving all the potential represented by love and by America and the American continent, and like all the greatest of dreams it couldn't possibly be achieved.

Trying to resurrect the chestnut is a dream too, one fueled, first of all, by the very human tendency to screw things up—to import, through sheer bad luck and the restless energy of trade, the bugs that destroy the chestnuts and then, decades later, the hemlocks. The dream is fueled also by the equally human impulse to set things right, to repair what's been broken. No, we can't fix everything just the way it was before, but maybe we can regain a bit of what we had in the past—not just for the sake of nostalgia, but because the world simply works better, is more whole and hale, when it has not lost so many pieces. Maybe Fred Hebard's and Lucille Griffin's great-grandchildren will be able to walk under tall chestnut trees, and pick up— *careful now!*—those prickly burred nuts, and taste a native chestnut stuffing in a nut-fed turkey. Maybe they will then recognize a bit of the richness America's pioneers found as they moved out into an untapped continent and sense the hard work their own more immediate ancestors put in to get some of it back.

If that green light does exist at the end of Daisy's dock, it's not in the shape of a single tree tucked away in some hidden Appalachian fold and waiting to be discovered. It's in the hearts of the people who are devoting themselves to working with the living tissue of plants, generation by generation, so that the trees yet to come can both reflect their original grandeur and manage to survive in the midst of our acquisitive, demanding, and endearing appetites.

2

Entering the Woods

*W*hen the settlers came west from the tumbled hills of Appalachia and the lush lowlands of the Ohio Valley they found something almost beyond imagining. They broke out of the shade and into the sun-struck openness of waving tallgrasses and sunflowers and great-leaved compass plants that lined themselves up with the North Star. It was a beautiful, frightful sight. Many of them were so cowed by the open space that they stayed along the river courses where the trees grew. They believed a place that couldn't grow trees couldn't possibly be fertile cropland, however thick its grasses and flower-spangled its summertime meadows. They were so flummoxed by this new land that, realizing they didn't have a proper word for it, they had to borrow from the French: *prairie*. The term meant something like *meadow* back in France, but here it was reborn, as so many immigrants were, and it came to stand for something entirely new: long, sloping rises and dipping swales and rolling plains of grass and wind and great weathers, a land of such breadth that it could take your breath away. During dry summers runaway fires could rampage for miles, and in winters the frigid wind down from the Canadian plains blew the snow sideways.

The prairies were harsh. But once John Deere figured out how to break the age-old soil, knit together with deep twining roots, they proved the perfect place for Thomas Jefferson's vision of a democracy of plain yeoman farmers, each family with its hundred and sixty acres and its freedom. For much of the nineteenth century the prairie was the place where the nation stopped facing east and plunged out, unhampered by terrain or history, into a destiny as clearly manifest as those Illinois and Iowa and Kansas horizons, a perfect orgy of righteous settlement that spread across the continent's midsection like spilled ink. Prairie was at once the wild and natural potential of the country and the nation's source of energy for turning that potential into something more practical; it was beauteous meadows and hard-working immigrant farmers in overalls; it was flocks of locusts and trampling herds of bison and the pastoral ideal of orderly fields and cattle lowing in green pastures. The prairie was the most American of American places.

The forest and the prairie—they were the yin and the yang of U.S. history writ on the land, the dark and the light, the old familiar terrain that harked back toward Europe and the new ground on which unprecedented chapters of history could be written. The forests became the home of rednecks and unreconstructed conservatives, the place where *backwoods* became equated with *backwards*; the prairies were the place where people looked forward, to the west, to the awakenings of the Progressive Era, to new possibilities. The dichotomy became a part of the American mythology. On the prairies, wrote Walt Whitman, American society would have a chance to grow "entirely western, fresh and limitless—altogether our own, without a trace or taste of Europe's soil, reminiscence, technical letter or spirit." "Of course it was possible—everything was possible out there," wrote Ole Rölvaag of prairie settlers in his novel *Giants in the Earth*. "There was no such thing as Impossible any more. The human race has not known such faith and self-confidence since history began."

It made a good story, anyway. Like any myth, the opposition of forest and prairie was an exaggeration, an overstatement that served a narrative purpose but skated blithely past nuance. There never was a great divide between the two ecosystems. Just consider the names of some towns scattered throughout the tallgrass prairie region: Funks Grove, Homewood, Hazel Dell, Scotch Grove, Elk Grove Village, Hickory Hills, Shadeland, Linn Grove, Oak Park Heights, Park Forest, Pleasant Grove, Oak Creek, Groveland, Hazel Run, Poplar Grove, Eagle Grove. These aren't names that would be given to wind-whipped prairie crossroads with scarcely a tree in sight. They're names that show where many of the prairie settlers put their roots: under trees surrounded by grasses. They're reminders of an

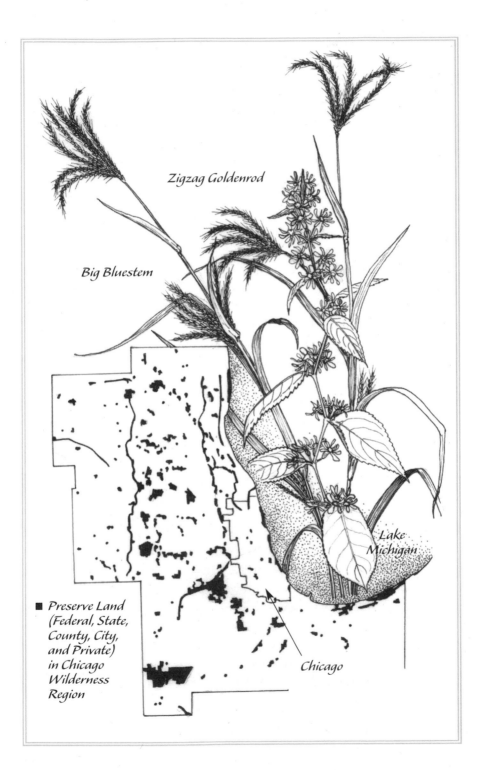

Zigzag Goldenrod

Big Bluestem

Lake Michigan

■ Preserve Land
(Federal, State,
County, City,
and Private)
in Chicago
Wilderness
Region

Chicago

ecosystem that by the late twentieth century had grown so endangered that its very existence had been almost forgotten: the midwestern savanna, neither forest nor prairie. How appropriate, then, that its rediscovery should have come about through the practice of restoration, which is neither *nature*, as that term is commonly understood, nor human, but a hybrid that combines the two into something new.

WHAT IMPRESSED STEVE PACKARD when he was a boy growing up in Shrewsbury, Massachusetts, were the huge, old hemlocks and white pines that grew in a ravine a healthy walk away from his family's house. Even well into adulthood he was not able to quite articulate their appeal, but there was something about them that he wanted a part of. He asked his father if he could dig up some of the little trees under the great ones and plant them in their backyard. His father said no, somebody owns that.

Nobody owns them, said little Steve; they're just there.

Somebody owns everything, his father told him, and then said, I'll tell you what, if you want to find out who owns them and get their okay, you can do it.

How could he do that? Steve asked. He learned that he had to go to the town hall. So he did—a small boy who couldn't see over the railing of the clerk's desk but insistently traced the roads on the maps to point the way to the ravine. He got the owner's name and called him up. The landowner was gruff, but gave in all the same. Just the little ones! he told Steve.

Steve's father was impressed at his son's initiative. Okay, good for you, he said. This is really good. Now I'm willing to go to the nursery and buy you trees of the kinds that you want.

I don't want nursery trees, Steve said. I want real trees!

The argument was fierce. In the end Steve's father drove to the ravine, and the two dug up a few seedling hemlocks and pines. Then Steve's father drove to the nursery. Steve set his jaw. His father bought a white pine that looked like it was on steroids. Back at home, they planted all the trees. The nursery pine was buff and vigorous. Next to it Steve's looked like Charlie Brown's Christmas tree.

Steve was never happier than when the nursery tree caught a disease, grew all gnarled, and stopped growing while his own persisted.

In adulthood Packard's tree planting gave way to tree felling. I saw him in action one winter Saturday when he helped me down an ash tree. It was a volunteer workday in December at Somme Woods, a county forest preserve about twenty miles north of Chicago. Remnant crusts of snow lay on the brown and tattered leaves underfoot. About fifteen volunteers had showed up to help Packard and his wife, Linda Masters, cut trees and brush. A brush pile the size of a Volkswagen Beetle had already been constructed in a small clearing. Our job was to expand the clearing so that sun-loving native plants, some of which had persisted in the modern shade, could thrive once more. I started to work on an ash tree six inches in diameter and was trying to cut it just above ground level. I had a dull bow saw—and more enthusiasm than smarts. The blade was soon stuck under the weight of the trunk.

Packard saw me sweating and struggling and came over with another saw. He is lanky and middle-aged, his dark hair silvering. He wore a beat-up canvas hat that looked as though it had perhaps once been white. "You're cutting that the hard way," he said. "Let's try this." Packard's voice is a bit reedy. He started sawing three feet above the ground, and by both of us pulling on the saw's opposing ends, we were able to fell the tree in less than two minutes. It toppled toward the brush pile, which a man named John McMartin, who was wearing dark green flame-resistant clothing, had been struggling to light. Slowly the flames caught. We cut the ash tree into manageable pieces and heaved them onto the pile.

The fire quickly engorged itself. It had no end of fuel; there was no end of other trees all around us. A few of them were big burr oaks whose trunks stood like great pillars, but the great majority were small and scruffy and no bigger around than the ash tree Packard had helped me cut. In every direction they stood in dense thickets, forming walls of trunks and branches that were hard to walk through.

We sawed and we lopped and we tossed the cut lengths onto the fire. We took off our jackets and sweaters. The flames climbed ten, fifteen feet high, and the outlines of the volunteers working on the

other side of the fire shimmered in the heat haze. When we threw
newly cut branches onto the pile they burst into flames even before
they came to rest atop the flaming mass. After an hour and a half it
was time for a break. We sat on downed logs and on jackets folded
on the ground and ate bagels, which Masters and Packard had
brought, and drank cool water. "What we're doing here is clearing
out the brush," Packard said. "You'll see there are big burr oaks
here, but few young burr oaks. They don't survive in the dark." This
woodland was once an oak savanna, he told us, where a rich mix of
grasses, wildflowers, and shrubs had grown amid scattered big trees:
burr oak, scarlet oak, shagbark hickory. Through decades of neglect
it had become choked with other plants, especially woody shrubs
and trees. Some of them, like the green ash I'd cut with Packard,
were native species; others, like the highly invasive European buckthorn, were not.

Once, when the restoration project here was young, Packard gave
a presentation about it to the local historical society. He described
how he and his volunteer colleagues were cutting out the brush and
replanting native species, then asked if there were any questions. There
was only silence. Everyone in the room looked at an older man sitting in the back. It was Louis Werhane, a scion of the family that had
owned the land before it became a forest preserve. The family had
farmed the land, Werhane said that day, until the Cook County Forest Preserve District kicked them off to make the preserve. After that
it had almost killed his father to see how the district had neglected
the place and had let all that brush grow up.

After that meeting the Northbrook Historical Society fully supported the restoration effort.

⁂

PACKARD IS STILL GIVEN TO SETTING HIS JAW and getting what he
wants. He is persuasive and he is committed. He is a celebrity in the
world of restoration, one of its best-known names. He is an activist
who has helped mobilize hundreds, even thousands, of people in the
Chicago area to spend their weekends, and their sweat equity, on
places previously neglected. He founded the influential prairie
restoration effort along the North Branch of the Chicago River in

Cook County's forest preserves in the 1970s, and two decades later he was a key player in the formation of the three-state Chicago Wilderness network, a groundbreaking effort to coordinate conservation and restoration work throughout the full breadth of an urbanized region. He has been a lightning rod for the controversies that have swirled around what has become one of the nation's largest grassroots restoration efforts. He has worked for a slew of conservation organizations and agencies; today he is employed by the National Audubon Society. His job description, at all the places he's worked, might consist of a single word: catalyst. Above all, he has through decades of practice become a dedicated advocate of the idea that it is volunteers—the regular folks who are tied to wild places simply because they live nearby and care about them and are willing to get their hands dirty—who should play leading roles in their conservation and management.

"Steve sure got the prairie fever," Robert Betz told me. Betz is a Chicago-area prairie enthusiast who became interested in those ecosystems in 1959 and became one of the strongest voices—and set of hands—working for their preservation. "He learned fast, never went to school to learn this stuff, but after he took the ball on prairie restoration the field really took off."

Like many conservationists, Packard loved nature as a child. He became a birdwatcher and went to a nature camp, but then came high school and college and a first career as an organizer against the Vietnam War. "Always I would escape to nature to be in touch with myself," he told me, but he wasn't persuaded that nature was something he should be doing. He thought he should be in the arts. He drifted. By the mid-1970s he was living in Chicago and working at a university library. One day he was at a friend's house and noticed a field guide to wildflowers.

"Can I borrow this?" he asked.

He threw himself into visiting Chicago's forest preserves and learning the flowers. He was meticulous and driven and kept precise notes. During his first year he learned how to identify 460 plant species.

Packard was a passionate bicyclist who found long, meandering routes to take through the county forest preserves, especially those along the North Branch of the Chicago River on the city's North Side

and nearby suburbs; he didn't have a car. That first field guide led him
to other books, and by the time he thought to stop at one of the overgrown forest preserve meadows just off his bike route, he'd been reading about the tallgrass prairie and knew which plants lived in them.

To his surprise, he found about a half dozen of those plants living
just off the trail. "This is it," he thought. "This is actually prairie!"

To the untrained eye, what Packard conceived of as patches of
prairie looked like old fields overgrown with brush and weeds. Cook
County, he had discovered, had a splendid system of forest preserves
that form a greenbelt around much of Chicago and its inner ring of
suburbs. When its establishment was proposed in the early twentieth
century, the forest preserve system was a visionary idea: it was to be
an antidote, both physical and psychological, to the densely packed
urban squalor of the city. "The woodlands," wrote Daniel Burnham
in his 1909 *Plan of Chicago*, which incorporated the forest preserve
idea, "should be brought within easy reach of all people, especially
the wage earners." Another of the Progressive Era planners behind
the forest preserves, the great landscape architect Jens Jensen, voiced
his support for the preservation of greenbelt lands in their "primeval
state . . . for the benefit of those desiring knowledge of the plants
indigenous there." It took until 1916 to get the politicking over with,
but then the county began to buy land with a vengeance. Any more
or less undeveloped land could become a forest preserve, whether it
was a woodlot, a rare unplowed prairie tract, or a farm.

The tracts in the most accessible places became picnic areas and
softball fields. Some of the meadows were mowed to provide dog-
walkers room to roam. Most of the preserve lands, though, which
came to total some sixty-eight thousand acres, or more than 10 percent of the county, were left to their own devices, partly because there
wasn't sufficient money to intensively manage them all, but also
because it was a reflection of the dominant ethos of land conservation
through much of the twentieth century. Leave the land alone, the
thinking went, and it will take care of itself. Grass and shrubs and
trees will grow as they should, and nature will see to it that the place
functions in a natural way.

The idea that the forest preserves were humming along nicely,
taking care of themselves, was a salve not only for the bureaucrats

charged with managing them but also for harried urbanites: your job might be a daily stress, the traffic might be getting ever worse, your neighborhood might be far too noisy or might be changing into something you didn't like, but out there in the forest preserves the world of nature was doing what it had always done. It was something to be proud of: one of America's most rough-and-tumble cities was allowing nature to persist and to be a part of its inhabitants' lives. It was good to know, even for those who never walked through the leafy green walls that lined the streets.

ॐ

IT WAS THE IDEA THAT NATURE would take care of itself, as much as any physical thickets of brush, that Steve Packard walked square into back in the summer of 1975 when he finally dismounted his bicycle and entered the woods. He recognized that the more open parts of the forest preserves supported lingering populations of prairie plants. He could also see that the places needed some work. Anyone could: trash and even burned-out cars lay strewn around the meadows. Packard's reading, though, had persuaded him that these relict prairies needed more than a litter cleanup. He knew that, left alone, the scarce prairie plants were slowly being starved of light and nutrients by shade-loving brush and trees, most of them nonnative species. The fragments were becoming woods, period—and not pleasant woods full of big trees, but dense stands of spindly trunks. They needed some clearing out. What they really needed, he supposed, was fire.

By the 1970s the idea that nature did just fine—regardless of what had been done to it before—when it was undisturbed by people was coming under attack in the scientific community, and nowhere more so than on the prairies of the upper Midwest. Of all North American ecosystems, prairies had, by the middle of the twentieth century, become among the most endangered. When French voyageurs first saw the Illinois River valley in the second half of the seventeenth century, at least a quarter-million square miles of tallgrass prairie covered an area that extended from the future state of Ohio to Kansas and from Minnesota to Missouri. By the middle of the twentieth century, only tiny fragments remained intact; in Illinois the prairie remnants strung along railroad tracks, in neglected rural cemeteries, and in vacant lots together were only one-one-hundredth of 1 percent of what had been.

The prairies were lost, in the main, to agriculture, as farmers discovered that soils that for millennia had grown big bluestem and Indian grass, white-fringed orchids and rattlesnake masters, gentians and goldenrods and myriad more species were rich and deep and ideally suited for corn and soybeans. They were lost to urbanization and the building of highways and cultivation of artificial woodlots. Yet even those few tracts that were by happenstance preserved came into decline. They might be spared the plow and the bulldozer, but if they weren't burned at least every once in a while, trees and shrubs grew into them, converting once-open grasslands into shrub fields or woods.

That prairies burned had been known since the first days of American settlement, when conflagrations that ran for miles across the open grasslands, often destroying outlier farms, formed a significant barrier to settlement. But only a handful of observers realized how ecologically important fires were until the 1930s, when the great naturalist Aldo Leopold proposed the planting of a prairie at the brand-new University of Wisconsin Arboretum in Madison. Leopold was a pioneer in the field of wildlife management, but in Wisconsin he also became an apostle of native landscapes. Because natural landscapes undisturbed by twentieth-century progress were becoming so rare in the Midwest, why not re-create some on the arboretum's two square miles for scientific study and public education? The arboretum, he said at the dedication ceremony in 1934, "may be regarded as a place where, in the course of time, we will build up an exhibit of what was, as well as an exhibit of what ought to be. It is with this dim vision of its future destiny that we have dedicated the greater part of the Arboretum to a reconstruction of original Wisconsin, rather than to a collection of imported trees."

Leopold's timing was good. Labor was abundantly available. It was the Great Depression, and two hundred young men from the Civilian Conservation Corps were living in tents on arboretum land. Some of the men had farm backgrounds, and to them the work must have seemed familiar. It was firmly set in the agricultural tradition.

They tried ripping bluegrass sod off the arboretum's old farm fields and replacing it with sod from relict prairies; they tried plowing parts of the fields and planting seeds of native big bluestem grass and other plants; they tried burning yet other tracts and scattering seed-bearing hay from native prairies on them.

Arboretum researchers carefully compared the success rates of the various techniques. They learned that although transplanting native sod appeared to be the single most effective means of jump-starting a prairie, it was labor intensive, and that other techniques worked too; that many of the native plant seeds needed a period of cold before they could germinate; and that too much success in growing native big bluestem and Indian grass could prevent many of the native wildflowers from growing. Above all, they learned that fire was the greatest ally of the native prairie species. Fire was the single most effective means of getting rid of bluegrass and other nonnative species that, unlike the natives, did not have deep roots from which new growth could resprout. Fire burned up the dead stems and leaves of big bluestem and Indian grass, freeing nitrogen for new plant growth and stimulating the deep roots to produce far more flowers than they could were the dead material not burned off. It was fire, too, that kept native woody shrubs and trees such as oaks and dogwoods from encroaching on the grasslands.

A graduate student named John Curtis had arrived at the University of Wisconsin the year the arboretum was dedicated. He soon became a botanist, a professor, and a colleague of Leopold's as well as a primary force behind the arboretum's restoration work. He became the coordinator of the prairie fires, which beginning in 1950 were set on a regular cycle; every three years, more or less, each part of what soon became known as the Curtis Prairie was burned. On those sixty acres Curtis showed that a prairie could be re-created—if you were willing to put in a lot of work and if you set a lot of fires. "The presence or absence of grassland is determined by the presence or absence of burning, with fire favoring grass and repressing trees," he concluded in his magisterial work, *The Vegetation of Wisconsin*, published in 1959.

Leopold died of a heart attack in 1948 while fighting a grass fire on a neighbor's land, and Curtis died in 1961, but the seeds they had planted bore fruit. The arboretum supported a thriving network of thirty plant communities representing what the upper Midwest had looked like before Europeans arrived. Other nature enthusiasts quite literally picked up the torch and began to use arboretum techniques— including fire—to create or to maintain prairies. In 1962, just as David Wingate was moving his young family to Nonsuch Island, botanists began a prairie reconstruction project at the Morton Arboretum, in Chicago's western suburbs. Bob Betz, who was a biochemist at Northeastern Illinois University but who explored the Midwest's remnant prairies at every opportunity, learned for himself that prairies needed to burn when he began finding prairie plants along railroad rights-of-way. Maintenance crews, he realized, burned those strips to keep brush from encroaching. To the shock of the farmers who lived near them, Betz began doing the same thing in the prairie relics that still clung to life in dozens of rural cemeteries. In 1972 he found a huge opportunity to put the idea of restoration into practice in the form of 650 acres of land inside the new accelerator ring at Fermilab, also on Chicago's western outskirts. The land had been in corn and soybeans. He used plows and combines, and a healthy dose of fire, to plant a prairie of natives there and to harvest their seeds. The site has since grown to more than twelve hundred acres.

It was no coincidence that the largest of prairie reconstruction efforts took place at arboreta; they were demonstrations, as Leopold indicated, facsimiles of the natural past of the North American continent—teaching tools. They looked pleasant and perhaps even wild to the casual observer, but to the knowing ecologist each was as self-consciously artificial as, say, the reconstructed colonial village at Williamsburg, where modern citizens could explore their cultural roots with the aid of professional interpreters wearing historically accurate clothing and speaking in carefully vetted accents.

In the late 1970s the Chicago Botanic Garden joined in. It had recently been located on a tract of reclaimed marshland in the north suburbs. Raw hills were bulldozed up from the winding course of the east branch of the Skokie River. Some traditional gardens were put in, full of herbs and roses and water lilies. Then the garden managers decided that they too should put in a prairie.

The garden happened to be just a mile from my family's house. I was a teenager looking for an Eagle Scout project. I volunteered. And so I came to spend a late winter and a spring watering flats in the greenhouse and, when the weather warmed, planting scores of compass plants and rattlesnake masters and prairie docks and other plants with equally evocative names. The seedlings, three or four or six inches tall, went into bare ground, plant after plant after plant. It was daunting: other volunteers and I emptied flat after flat of new plants, and when we looked up, the prairie-yet-to-be stretched on and on, going up a long rise that culminated in a hill overlooking the nearby Edens Expressway. How was all that going to be planted? I wondered.

I earned my Eagle badge that spring and went on to other things, and the prairie went on too. The garden staff and volunteers continued to add to it. It came to occupy about fifteen acres, and today it is a beautiful sight in late summer, with the flowers blooming and the grasses ripening to bronze and small flocks of goldfinches jauntily perching on stems and looking for seeds. The wind sweeps through the tall heads of the grasses, unseen insects make little ticking sounds, and the traffic zooms by. The prairie occasionally gets burned in spring, but the fires are set by hand, not lightning, and are closely watched. They bear little resemblance to the raging, miles-wide prairie fires I'd read about. The prairie, for that matter, is far from a simulacrum of the marshy backwaters that had once existed on that spot.

That was how I got to know prairie, as a sort of museum piece built where no prairie had been before. It was something to be carefully planted and tended. It was something you got to see as part of the tram tour after paying an admission fee. It was a garden, and it had little or nothing to do with the wild and unkempt forest preserves, like the dense tangles in Somme Woods. Those preserves were the places where, as an adolescent boy, you could go to light fires, smoke cigarettes, drink illicit beers, get half lost, and engage in all the other timeless pursuits of late childhood. In those thickets big puddles of mud formed after rains, and the thorny branches of the small trees snapped back in your face if you were following too closely behind your buddy. You had to duck and sidle and squeeze through to get in. It was those places that were *nature*—those places, rather than the tidy prairie, that seemed to me the way the world must have been when it was given.

❧

BUILDING A PRAIRIE MORE OR LESS FROM SCRATCH, then, had prece-
dent by the time Packard made his discoveries along the North
Branch; before too long, he'd identified seven prairie scraps there.
Restoring an existing but degraded prairie in a wild setting, though—
well, that was pretty much uncharted terrain. It was the difference
between building a new church in the style of a Gothic cathedral but
with modern technology and restoring an original cathedral that had
been modified by centuries of ill-thought-out additions and alter-
ations. A good deal of sensitivity, Packard realized, would be required.

"I knew how to organize things from the sixties," he said, "and I
said to myself, 'I could do this. I could save these prairies. But if I do
save them I'm getting myself into something for the long term. I'll get
committed to it. I'll bond with them. I won't want to let the thing
fail. I'll end up with ties to a lot of people. It will tie me. Am I going
to do that or not?' And just one time this little dam broke, and I
decided, 'I'm going to do it.' It was a feeling not terribly different
from deciding to get married, or have a kid. You can see the future:
there's going to be travail, and many good things."

The future that Packard saw was of lively bunches of ecologically
literate volunteers working in the woods every weekend, planting
native species, cutting the brush that was threatening to overwhelm
the sun-loving natives, setting the fires that would maintain and
enlarge the remnant prairies. The preserves in their ecological structure and species diversity would, over time, approach the same condition they would have been in had the natural cycle of prairie fires
never stopped in the nineteenth century. They would continue to
evolve as natural prairie tracts would have, had there been any left.
And if human beings were required to pull weeds and set fires into
the indefinite future to maintain these little tracts, that was fine too;
modern people were, after all, as much a part of the ecosystem as the
Native Americans had been.

He was going to do it. That was the decision that made all the difference to Steve Packard, and to the North Branch prairie remnants.
It was the end of his period of drifting. In August of 1977 Packard
went to a Sierra Club meeting and announced that he was starting a
restoration project to be called the North Branch Prairie Project and
that work on it was commencing that very weekend. The blooms of

smooth phlox—a lovely prairie wildflower—were ripening at Somme Prairie Grove, and Packard wanted to collect them so that they could be planted elsewhere.

He led a dozen new volunteers in doing exactly that. They collected two thousand seeds. A week later he led his volunteers in planting the seeds, one by one, at three other North Branch prairie remnants. It was tedious, time-consuming work, and Packard knew it wouldn't bear results for a long time. He wanted to ensure that volunteers who might not share his heartfelt love of obscure plants, or his confidence in his long-term vision, had reason to come back week after week. They needed some results they could see. So that same summer he told his new volunteer corps: we're also going to cut brush so that we can expand the prairies. We're going to cut the buckthorn and the gray dogwood and the green ash. Once we've cleared an area out we'll plant the native wildflowers and grasses there and watch the tiny, beleaguered prairie fragments expand and grow.

Some prairie experts expressed reservations about Packard's idea of just scattering the native seeds on the ground. They were, after all, used to the agricultural model of plowing the ground and then planting native tallgrasses and forbs like so much corn, or the highly labor-intensive transplanting of seedlings I'd experienced at the Botanic Garden. At their suggestion, Packard did try a few of those techniques. He borrowed greenhouse space and raised some seedlings, but it was simply too difficult to water them once they were in the ground. At the suggestion of one volunteer, he led his helpers in cutting small squares of turf out of the restoration areas and planting seeds in the newly exposed bare ground.

"That was a very important experiment," Packard said, "in part because everything we planted in there died. Conditions were much harsher. The sun would bake down, the other plants would put their roots in; it would dry out to nothing." He kept watching the plots, though, just because he was curious, and he eventually came to notice that some of the planted natives were surviving. "They were living around the very edges of these places," he said. "They were living where they were competing, and in a niche created by other plants. I said, 'Okay, we're just putting seeds in the turf, then. We're not digging things up anymore.'" Many prairie plants, it turned out, thrive on competition. As long as they're not shaded by trees, they grow just fine when densely surrounded by other grasses and forbs.

Between natural history lessons, Packard was getting his own lessons in politics. A year after the project started, forest preserve maintenance workers mowed all the places the volunteers had been working on. Some of the volunteers were dismayed, but Packard wasn't; he had his eye firmly on the long view. He coordinated a protest to the Forest Preserve District. Eminent scientists wrote letters; volunteers who lived near one of the restoration sites in Chicago, the Sauganash Prairie, complained to their ward committeeman—a political figure who in Cook County politics carries a lot of weight.

The mowing didn't happen again. But suddenly Packard faced another challenge. In the summer of 1978 he got a call from one of the forest preserve staff: Steve, the man said, let's visit these places together with a few of the other staff members. We want to see what you're doing.

Packard agreed. He was early to the appointment and so were the preserve staff members, but they didn't want to start early. They seemed to be waiting for something. Packard didn't know what. Then another car drove up.

"A guy with a beard got out and starting walking over to us," he recalled, "and I said, 'Uh-oh.' And then they introduced me to Dr. Betz, and they said, 'We have Dr. Betz here to tell us whether these are really prairies or not.'"

Together the group visited three of the North Branch restoration sites. Packard had his pitch all worked out. Of course we know we only have *some* prairie plants, he said. They might not be prairies now, but they will be in the future. Isn't it nice that people care about them and that the neighbors are getting involved? Furthermore, if you let us work on these trial prairies, then people will visit them and won't go tromping through the real prairies like the ones Dr. Betz studies.

At each site the forest preserve staffers asked Betz what he thought. It was a bit like asking something of the oracle at Delphi; each time he replied evasively. Finally, after the third site he said, "Well, I've seen enough, and I can tell you these are *incipient* prairies."

Packard had won that battle. Betz, though, had some advice for him. "Don't keep pushing so hard," he whispered to Packard at the end of that day. "You've won. You've won. Don't keep pushing so hard."

Betz came to be one of Packard's mentors and readily shared his expertise with the younger man. One of the things Betz taught Packard was the importance of political maneuvering. Betz had already spent more than a decade exploring the prairie fragments of Illinois and watching them decline; he knew firsthand that they needed maintenance. He was, in fact, given to scattering seeds, just as the North Branchers came to do, so as to increase both the genetic pool of the rare plants and the diversity of remnant prairies. He also knew, however, the mythic power attached to the notion that nature can best take care of itself when left alone. He worried that politicians and members of the public alike would attach less significance to the remnants if they knew that he or others were messing with them: if a prairie could be planted, why preserve the few fragile relics that were left?

Once Packard attended a field tour Betz was giving at Markham Prairie, one of the choicest prairie fragments in the Chicago area. Packard knew that Betz had planted a lot of species there. During the tour he asked him, trying to put as much innocence in his voice as he'd had as a boy asking for tree seedlings, "Dr. Betz, you're talking about this needing a lot of help to get back. Would a person ever plant a plant in a place like this?"

Betz looked at Packard, looked at the crowd, and boomed, "NO! It's nature! Leave it alone!"

ॐ

BUT LEAVING IT ALONE WAS AN IDEA from which Packard and his fellow North Branchers began deviating more and more. Being Packard, he did keep pushing the officials of the Forest Preserve District. The next thing he pushed the forest preserve officials on was that they had to burn the prairie patches; otherwise all their work in clearing brush and scattering seeds would be in vain.

This statement was not hyperbole. Within a year of the project's inception Packard and his volunteers could see how quickly the brushy species grew back in after their trunks had been lopped off.

Applying herbicide to the cut stumps kept them from resprouting, but even so, new plants were quick to emerge. Without fire, Packard knew, the North Branch Prairie Project would be caught in a continual holding action, eternally cutting back the woody vegetation that kept growing, Hydra-like, into what ought to be prairies.

The idea encountered surprisingly little resistance. The superintendent of the forest preserves knew just enough about the role of fire in the Wisconsin and Illinois prairie restoration projects. "He sort of knew that what we were doing was in the spirit of the forest preserves as he understood it," Packard said, "and he knew that it was needed, and that there was no way they [the forest preserve staff] were going to do it. He wanted to keep it down so that it wouldn't become a political problem for him, he wouldn't have to spend a lot of time on it."

The areas being talked about were small and generally out of the way anyway. The forest preserve officials told Packard yes, you can set fires, but we'll need some of our staff there, along with a fire engine in case things get out of hand. The early burns proceeded with excruciating slowness. Packard and company envisioned blazes something like the illustrations they'd seen of early prairie fires, with stampeding bison fleeing a twenty-foot wall of flame. Now *that*, they figured, would take care of the buckthorn. That, however, was an image to give heartburn to a public land manager, especially one in charge of urban preserves lined by neighborhoods. The forest preserve staff tended toward caution. They were wont to set tiny "backfires" at the downwind side of the area to be burned; they'd travel a few feet and go out. Then, often as not, the fire engine would get stuck in the springtime mud and would need to be towed out. Progress was frustratingly slow.

"They liked to do these little strip burns where they'd back up twenty feet and light the fire again, so that it took them all day to burn three acres," John Balaban told me. He and his wife, Jane, were among Packard's early recruits on the North Branch and continue to remain very active there today. They, like Packard, experienced a great sense of relief when a forest preserve official told Packard, two years into the burn program, that they were too busy, that Packard could go ahead and conduct the burns without them. And without a fire engine.

The first big burn came in April 1982 at a preserve called Miami Woods. The restorationists lit backfires along the prairie's downwind side, then along the two sides adjacent to that. When those narrow swaths were blackened they lit the upwind side. The flames raged downwind through the tall grasses. Had bison been there, they would have fled. Then the flames crashed into the blackened edge and went out, just as had been planned.

It was also in 1982 that those first seeds Packard and company had scattered came up—the smooth phlox collected on the first workday back in 1977. The successful fires, and the rebirth of the phlox and of many other species that were taking on new life, were an inspiration to the North Branchers. Throughout the old prairie fragments, they saw the big bluestem and the golden alexanders and the rough blazing stars coming up and thriving and blooming. Where they cut brush and burned they saw the buckthorn and the bluegrass and the dogwood knocked back and the open grassland they'd read about in historic accounts reviving itself.

We must be doing this right, they thought, carving the old prairie out of the new woods. The land itself was showing them—and the Forest Preserve District, and anyone else who cared to look—that their template for it was correct.

And it wasn't just plants. In the late 1970s Packard first saw evidence that white-tailed deer were coming back. They'd been hunted out of the state by the beginning of the twentieth century and were making a slow comeback. "It was wonderful to see those first footprints," he recalled when I interviewed him for a magazine story in 1992. "We thought it was wonderful to have another bit of the original fauna back." At Somme Prairie Grove he found a tree stand where poachers were illegally staking out deer. He was angry about that, in those early days, and felt like reporting the anonymous hunters to the authorities.

The life of the North Branch Prairie Project settled into a seasonal rhythm. Burn in early spring, just as the bluegrass was coming up. Follow that with a liberal scattering of last year's collected seeds.

Spend the summer cutting brush and harvesting more native seeds. Where trees were too large to easily fell, kill them instead by girdling them, removing a strip of bark around their entire circumference. In the late fall, conduct a seed party and collate the entire year's haul of seeds into mixes for different conditions: dry prairie, wet prairie, something in between. Winter: more brush cutting and burning the year's accumulated brush piles. And throughout, during each work-day, allot time to eat bagels—always donated by the volunteer in charge of the particular site—and talk about the reasons for doing what was being done and about their shared visions for the land.

The volunteers were pioneers, and although many didn't last long, many others did. They grew eager to see what each new year would bring. "What restoration does is—we like to talk about own-ership, but I think what happens is the land owns us," John Balaban told me. "You get so attached to a piece of land that you have this need to go back there to see how it's doing, and this anticipation: I can't wait until the next burn season, and then I can't wait until the next spring season after the next burn season, because I want to see how it's improving, how it's getting healthy, et cetera."

And those bagels, punctuating the hard labors of workday after workday, were pretty good too.

⁂

WHAT IMPRESSED RICK SIMKIN, on one of his first outings in the North Branch preserves, was seeing that a twelve-year-old boy was among the volunteers. Simkin is a slender, bearded computer pro-grammer whom I met working that winter day out in Somme Woods. "He was a very withdrawn kid," he told me, "but somebody handed him a saw and told him to cut down a tree. It was an adult job, and it was probably the first time he'd ever been given an adult responsibility. He did a good job at it."

Simkin himself came to the project because he felt a need to take responsibility. He was at a party once decrying the environmental problems that were wracking the planet. "Well, what are you doing about it?" someone asked him, and Simkin realized that all he was doing was worrying. It was not sufficient. Since then he's made a point of coming to the woods periodically to clear brush.

By now, thousands of people have done the same, and if some have signed up only for the opportunity to do occasional feel-good grunt work such as cutting down small trees, many others have signed on for the long haul and have become leaders and decision makers in their own right. Packard likes to compare the restoration work to other long-term projects that exceed a single lifetime. "I've told people this is like building a cathedral," he said once as we walked through Somme Prairie Grove. "Your children's children are going to be working on this. The instant gratification crew goes elsewhere. Older people seem to understand it better. And you need people with expertise in stonework, glasswork, and so on. No one person has all the expertise. We need people who know the plants, the frogs, who know how to conduct prescribed burns. We need the artisans for the cathedral."

Training the artisans quickly became as much a part of Packard's mission as cutting brush or setting fires or massaging the Forest Preserve District. He knew from the beginning that he would not be able to oversee all the restoration work; the North Branch preserves alone were too many, too widespread, and too diverse, and beyond them were many more acres of preserves in Cook County, and beyond them tens of thousands of acres in the other counties of the Chicago area. He began recruiting other enthusiasts to serve as "site stewards" for preserves. Among the first were the Balabans, who moved to the North Branch area in 1979. They'd previously met Packard at a prairie on the south side, and when he found out they'd moved north he told them, "I've got something for you guys!" In 1980 they became the stewards of the Bunker Hill Prairie in Chicago. In 1986 they adopted a second site, Harms Woods.

I talked to them one summer evening at their tidy house in suburban Skokie. We were, audibly, under the flight path to O'Hare. John and Jane are solid people, middle class. You could walk past them in the mall or in the Little League stands and have no idea how they spend their weekends. They soon became two of the resident North Branch experts on plant identification. They still like nothing better than keying in on some obscure and difficult-to-identify plant family. They learned where to look for and how to identify sedges and knotweeds and many other tricky plants. Studying the preserves became a great treasure hunt that opened into ever-increasing levels of complexity.

"When you first go out you notice the lilies and the orchids and the big things that you can't miss," said John, "and then once you identify them and get to know them and are comfortable with them, then you notice the smaller things, the asters or the fleabanes, and now it's not unusual for people to ask us to lead a sedge class. The flowers are green and it looks like grass and you need to key them out to tell them apart! So there are always more and more subtleties, and more and more things that you can get into, learning about the site and how the pieces fit together."

Unlike Packard, the Balabans didn't come to make a living of restoration. Jane is a retired pharmacist and John is a high school math and physics teacher. "You talk about restoration taking over my life," he said, "but it really is secondary to that. People ask me what I am or what I do, and I talk about high school." Yet it is a love of learning that clearly bridges their lives. Like David Wingate, they seemed to me to be able to keep in their minds both the big picture—the overarching vision that gives purpose to the daily grind of what amounts to some very hard work—as well as the smallest of details. It seemed a nicely holistic view to me.

"You put seeds out of native species," Jane said, "and all of a sudden you first see something come up whose seed you've been putting out for ten years, or you see a plant that is simply new to you. We've become interested in dragonflies, and just this past summer we found a whole new dragonfly and got pictures of it. It's a constant learning experience watching how the land responds to management, how different years or the weather affect things. It's constantly changing."

✣

THE NORTH BRANCHERS' CURIOSITY about how the land works came to prove useful in the early 1980s when they began working to reestablish prairie plants around a stand of big burr oak trees on the edge of Somme Prairie Grove. The oaks were impressive: sixty and more feet tall, with foot-thick branches and massive gnarled trunks. But they'd become surrounded by great crowds of smaller trees, both nonnative buckthorn and honeysuckle and native species. "Where the oaks were close, we could see their heavy, twisted black limbs through the thin gray branches of ash and aspen," Packard later wrote in an article for a new journal called *Restoration & Management Notes*, published at the University of Wisconsin Arboretum. "It was as if they were in prison or refugee camps, their lower limbs dead or dying in the deepening shade. 'Free the oaks!' we sometimes joked."

It was well known that isolated oaks with thick, fire-resistant bark had grown in places on the presettlement prairie, and Packard imagined taking the thickets out and seeing the prairie tallgrasses and wildflowers forming a continuous carpet under these particular grand oaks. He had his crew cut out some of the brush. Prairie species didn't return; the brush did. They tried setting a fire under the oaks. Low on fuel, the fire poked and curled its way through the downed oak and buckthorn leaves, but its flames rose only four or five inches and never seemed to come alive as fires did on the open prairie plots. Still, it did succeed in killing some of the brush. The North Branchers' hopes soared with the increased visibility in the grove. Then a bunch of invasive thistles came up in place of the brush. It was not the flowery prairie with oaks that Packard had imagined. At Miami Woods, where the volunteers also burned under oaks and sowed prairie seeds, rich grasses came up, but Packard didn't recognize them. In the fall of 1984 he collected some specimens and put them away, thinking that something was wrong. The sites weren't reacting according to the prairie pattern.

That winter he spent a good deal of time reading up on botany. He combed the eight-hundred-page *Plants of the Chicago Region* from cover to cover. He wasn't trying to learn individual plants as much as he was looking for patterns. The book allowed him to do so because it mentioned the other species with which any given plant was typically associated. Packard keyed out the Miami Woods grasses and found that they, and some of the other species that had come up near the oaks, were associates of one another. *They're not prairie species at all, he thought; they're something else.* Packard was working for The Nature Conservancy by now, and he thought back on a memo the Illinois office had received from a national conservation planner in the summer of 1984. "Savannas are nearly exterminated, everywhere in the Midwest," it read. "Remnants should be saved and restored."

Maybe, Packard thought, the oaks at Somme and Miami Woods hadn't simply towered over a bunch of prairie plants. Maybe they were markers of an ecosystem of their own—something unique and, according to The Nature Conservancy, critically endangered. He began listing all the plants that had cropped up at those sites. Then he went back to the library. He began perusing other plant lists, some of them quite old, looking for species associated with midwestern woodland edges and oak openings. Eventually, he compiled a list of 122 plants that he presented as "distinctive savanna species" at a prairie conference in 1985. That year the North Branchers began collecting their seeds wherever they could be found, often along unmaintained woodland edges.

Packard's paper caused a stir among the prairie aficionados. What he was doing, many felt, was lobbing a missile at the revered memory of John Curtis, the Wisconsin botanist who had done so much to make prairie restoration possible. In *The Vegetation of Wisconsin* Curtis had recognized the existence of tallgrass-oak savannas, but he had claimed that they didn't harbor any distinctive species; rather, he thought that they were merely transition zones between woods and open grassland. Now here was Packard, a firebrand with no formal botanical training, claiming to have discovered a savanna community largely through processes of field experimentation and apparently random prospecting in the ecological literature. The experts were dubious. When Packard showed the paper to the editor of one professional journal, he responded that the approach was "dangerous" and suggested that Packard leave savannas alone.

Some practitioners, though, found Packard's list profoundly useful. One biologist from McHenry County, just to the northwest, read it and found himself remembering a weedy site that he'd evaluated for its biological potential years earlier. He hadn't thought much of the place, but now he realized that many of Packard's species grew there. Packard went for a look, and it quickly became one of the county's top priorities for acquisition as a nature preserve. Today the Middlefork Savanna is regarded as one of the highest-quality savannas in Illinois.

It was not for another year, though, that Packard was able to put the capstone on his theory. He was able to find an obscure article published back in 1846 by an early Illinois doctor named S. B. Mead. Like many physicians at the time, Mead had a broad interest in natural history, and he compiled a list of the plants he identified in his travels around frontier Illinois. That was interesting, but what Packard found compelling was that Mead had included abbreviated habitat notations with each species name. P meant prairie, T timber, W wet, H hills, and S sand. B stood for "barrens," an early term for rather open areas pocked with scattered stands of trees. Packard tingled. The penny dropped. Barrens were savannas! One hundred and eight species were marked with a B, and once he'd been able to translate their names into modern taxonomy he realized that there was a marvelous overlap with his own list. He was vindicated. In fact, the plants that he had hypothesized as belonging to and making up the oak savanna had, a century and a half earlier, been spotted exactly there—and not, for the most part, in the neighboring prairies or woodlands.

"I felt as though I'd found a Rosetta Stone for the savanna," Packard wrote. The savanna was real in the historical literature; more important, it was real at both Somme Prairie Grove and at Miami Woods, where in the spring of 1986 dozens of savanna species were emerging: silky wild rye, bottlebrush grass, starry campion, blue-stemmed goldenrod. They were outcompeting the thistles and the buckthorn. When in 1989 a pair of eastern bluebirds—a rare species in Cook County at the time—nested at Somme, Packard and his colleagues considered it a victory both for their understanding of the savanna and for their technique: learning by doing and not getting hung up on theory.

Learning by doing—or at least recommending it to others—could drive some ecologists around the bend, however. The controversy about whether the savanna was its own ecosystem or a hybrid between woods and prairie went on. Some ecologists seemed to object to Packard's ideas all the more vehemently as they gained in popularity. One of them was Jon Mendelson, a biologist at Governors State University. In 1992 he, with two coauthors, published a paper sharply critical of Packard. Mendelson and his colleagues claimed that Packard's savanna template was being applied uncritically at other Chicago-area sites that ought to be woods, and that the savanna didn't really exist at all. "The management activities so popular today . . . are not so much noble attempts to restore a unique savanna community as they are increasingly destructive efforts to shove existing assemblages of species in one direction or another along a vegetational continuum," they wrote. "The end product is not the limitless creativity of nature, but a form of landscape architecture limited by the imagination of the manager. . . . Insistence on a preferred archetype is merely placing nature in a historical role and relegating its existence to that of an artifact, encased behind museum walls." The work of the North Branch volunteers, in this view, was not much unlike the explicitly artificial re-creation of ecosystems at the University of Wisconsin Arboretum.

Critiques of ecological restoration were gaining currency in the late 1980s and early 1990s, although perhaps less prominently in the world of ecologists than in the world of environmental philosophers. The first salvo came from the Australian philosopher Robert Elliot, who in 1982 published an influential paper entitled "Faking Nature" in a philosophical journal. Any restoration of nature, he argued, was less valuable than the original; in fact, a restoration of a wild place was akin to a forged copy of an original artwork. "Knowing that the forest is not a naturally evolved forest," he wrote, "causes me . . . to perceive the forest differently and to assign it less value than naturally evolved forests."

It was not a coincidence, perhaps, that such an argument came from Australia, which in the early 1980s was a far wilder place than Chicago and where conflicts were erupting regularly about the damming of virgin rivers, the logging of old-growth forests, and large-scale mining. In such a context it made sense to argue that honing the ability to bring places back from damage might excuse causing the damage in the first place. Chicago was a very different place, though. Its forest preserves were extensive and might look virginal to the untutored eye, but to Packard and his colleagues they looked profoundly damaged. To argue that they should be left alone, they believed, was tantamount to suggesting that cancer patients should be encouraged to heal themselves without the benefits of modern medicine.

"Some people don't, but I like the metaphor of the comparison between ecosystem health and human health," Packard told me. "There are a lot of diseases that doctors knew how to cure before they knew what caused them. They considered it their job to cure people to make them healthy again. The more you know that this disease is caused by a parasite; this disease is caused by a bacterium; this disease is caused by a vitamin deficiency—the more you know that, the better a job you can do. But if you're a doctor, what you most want to do is this: 'If I give this person a lime juice, he gets better. I have no idea why, but I recognize these symptoms, and I'm going to give him lime juice.' A lot of ecological restoration is giving lime juice to the ecosystem. It would be better if you knew why, but you can recognize health like you can with a person. If things are flowering and making seed and new plants are growing and the rare birds somehow like this place better than other places; when you study the snails you find that all kinds of rare snails are showing up here or are common here; they're dying out in other places—that seems like ecosystem health. It makes me happy. So I try and do that. And someone says, 'Well, what about this theory? What about this philosophical concept?' I'm interested in those things, but mostly they don't tell me much.

"It's easy to come up with cute ideas that do well in philosophical arguments that are irresponsible so far as the development of a healthy culture that works for people in the world. I don't spend too much time thinking about them."

A more immediate concern, though, did present itself in the form of white-tailed deer. Within only a couple of decades of their arrival in suburban Chicago their population had swelled to such an extent that there were more car–deer accidents in Cook than in any other Illinois county. Municipal and forest preserve officials in the region watched as populations of spring wildflowers, and expensive landscaping in suburban gardens, were eaten to oblivion; in one northern Cook County preserve twenty-six species disappeared within about a decade. When some of those officials suggested shooting deer to save wildflowers, they received death threats from animal-rights advocates who claimed that there was no overpopulation problem or, if there was, that the deer should be removed by nonlethal methods. At one point Packard chaired an advisory committee addressing the issue in Lake County, which adjoins Cook to the north. It was very difficult to find any common ground.

"The deer people, the protestors, were so ferocious," he recalled. "They'd go after people—like *seriously* go after them. People who thought they were dedicated to their work would quit and move away. They would say to me, 'You can't stand up to these people. They go after your wife. They go after your kids. The kids at school are mean to our kids.'"

Slowly, conservationists and preserve officials were able to make some headway in controlling deer numbers, but bad blood remained. At Somme Prairie Grove, Packard came to see firsthand the damage deer were doing. Lilies and orchids were disappearing. The solution to the problem seemed obvious. "People have killed deer as long as there has been this ecosystem here," he said. "That's been a substantial part of what made the ecosystem function. It's not like we're making it less natural by killing the deer. We're making it *more* natural. It's more like nature has been here if people are killing deer than if they aren't. To me it's a pathological alienation that says people having anything to do with something makes that thing worse, and corrupted. To me that's a philosophical problem, not the idea that people have a role to play. The human species are not aliens on this planet." Where the poachers' deer stand stood he began cleaning up beer cans and other garbage, hoping that their presence wouldn't be noticed by others.

⁓

ONE EARLY SEPTEMBER AFTERNOON IN 2003, fourteen years after the first bluebirds appeared, I attended a volunteer workday at Somme Prairie Grove. It was a sultry Sunday, the day before Labor Day, and only about ten people showed up. The newsletter announcing the workday's location had gone out a bit late. We introduced ourselves in the gravel parking lot. Four lanes of traffic pounded by. The oldest volunteers were retirees, the youngest fresh out of college. Some had been coming for years, and some were here for the first time. Then we followed Packard, single file, on a narrow trail through the tall grasses of the prairie. The sun-burnished seed heads of the big bluestem and Indian grass reached well over my head, and I found myself squeezing my eyes shut, repeatedly, to avoid being poked.

After a few hundred yards we reached the edge of the grove. Cicadas droned above us. Jets scudded low overhead on their way to O'Hare. The deeply corrugated trunks of the largest oaks were a good two feet or more in diameter: beautiful, sturdy trees. This savanna, though, was a ghostly and camouflaged one that was largely hidden behind other plants. Around us was a chaos of shrubby vegetation choking out the grasses and wildflowers. Some brush piles had been started, but they were surrounded by legions of head-high new shrubs: prickly buckthorn, smooth dogwood. Our job was to cut them and throw them on the piles. It's even easier lopping off a slim buckthorn than it is planting a seedling from a flat, but as I looked around at the other volunteers half-hidden among the leaves I had that same feeling, as so long ago at the Chicago Botanic Garden, of being overwhelmed by the extent of what needed to be done. I understood, viscerally, why fire was necessary to keep this landscape open.

Packard came around with a yard-long white plastic pipe. It had a big swab on one end, which was dyed cherry red with herbicide. He used it to daub all the little stumps we left behind so that the roots wouldn't resprout.

Many volunteers, Packard said, dislike the use of herbicide when they begin work. They want the purity of not relying on it. Packard always challenges them to prove that restoration can be done and maintained without their use.

"Without exception they give up or begin using herbicide," he said. "Without exception."

He moved on, and after an hour he called us together to go elsewhere. My shirt was soaked with sweat, my arms abraded with tiny cuts and scrapes. We walked on the narrow dirt trail through an area that had been cleared earlier that summer. Now it was a minefield of tiny buckthorn stumps a few inches high. Low goldenrods and wild leeks—and even tinier buckthorn seedlings—grew among them, but much of the ground was bare. Then we entered the old grove that had been cleared years ago.

It was beautiful. Oak trunks thick as architectural columns rose straight up from an understory rich with wildflowers and grasses. It was lush and green and open, and I reflected on how savanna restoration has no doubt been bolstered because it results in landscapes that many people really, really like: open but shaded, expansive yet intimate. It's no coincidence that the most expensive neighborhoods in the tony suburbs around Somme Prairie Grove are those that look the most like savannas, with stately oaks and lush lawns and well-spaced beds of shrubs. People simply like that sort of setting. It's an advantage of which people working to restore other landscapes may not be able to avail themselves.

We ate our bagels and drank our water around a big downed log. Packard had planned to have us work here, but he'd been looking around. Now he spoke up: there were too many rare plants— wide-leaved panic grass, false dragonheads, zigzag goldenrods—for us to work here in the summer. We would trample too many of them. Tackling the buckthorn around the edges of the existing clearing would have to wait until fall or winter.

We trouped on out of the grove and through the prairie again, where I was learning to keep my eyes open, then into a grove of trees near the parking lot. Packard pointed at a dense patch of skinny-trunked buckthorns ten or fifteen feet tall. Nothing else appeared to be growing amid them; the soil around them was bare or covered with a few drifted dead leaves. "This was a prairie in my recent memory," he said, "and now it's death." It was a tract that hadn't been cleared or burned, he said.

"People think *nature* means 'leave it alone,'" he went on. "But it doesn't. Etymologically it means 'it reproduces itself.' The average person wouldn't say, 'Oh, this is terrible,' but this is the fate of unmanaged land. The buckthorn kills everything underneath. Because the grass and wildflowers are gone, the soil is washing away, and the ponds the frogs and salamanders need to breed in are filling in. It's like a person getting old—the systems start not to work. But an ecosystem is supposed to be immortal. It's supposed to persist and keep going."

With that he had us going again with the saws and loppers. The small trees came down easily, and we dragged them off toward the road, working like archaeologists to expose something that had been hidden for a long time.

ॐ

THE NORTH BRANCH PRAIRIE and savanna fragments came to grow larger and larger with the progressive brush cutting and regular fires. So did the network of people tending them. In the 1970s the restoration work was confined to a few small preserves. In 1983 Packard and the other project leaders focused the project's growing energy by developing what they called the Volunteer Stewardship Network. The idea was that each preserve would have a designated steward in charge of organizing workdays and making decisions about what sort of work should be done in which places. The stewards would, in turn, support one another in their work. They would visit one another's preserves and, trading in the currency of ecology, learn from and rely on one another's expertise.

By the mid-1990s, the region's conservation leaders believed that it was time to ramp up once again. By then the North Branch volunteers numbered in the hundreds and were working at sixteen separate sites. They weren't just prairies; some were oak savannas, some were woodlands and forests. Other local groups had cropped up to emulate their work. Throughout Cook County forty-eight sites were undergoing volunteer restoration work. The Forest Preserve District had hired a land manager to oversee restoration plans and was budgeting $75,000 a year to buy tools, seeds, and other supplies for the volunteer effort. The district superintendent assessed the value of the restoration work the volunteers did at more than five times that amount. And the restoration ethic didn't stop at the county line: across the entire Chicago metropolitan area more than five thousand volunteers were annually contributing what was estimated at more than sixty thousand hours of labor. Restoration was going mainstream.

One of the newer volunteers was Karen Rodriguez, who'd worked with people who were mentally handicapped and then switched to a career in visual communications. She became active in the North Branch work in 1989, and her passion for restoration work led her to a third, new career working for the federal Environmental Protection Agency (EPA) in Chicago. She now coordinates grants for ecological restoration projects in the Great Lakes region for the EPA.

"When I started it was definitely individual little groups, each doing their own thing," she told me at her office in a downtown skyscraper, "but very quickly that evolved into dragging other agencies and groups in to help out. The work has not changed, but the social construct of the North Branch has become much more complex. Now it's much fuller. The conversation isn't just about how to cut brush or what herbicide to use. Now it's about art and restoration and more regional topics. The volunteers that are there are still doing the same work, but the scope of what they see as their purview as volunteers has become a regional view. They've come into a broader perspective that we didn't have earlier."

That new perspective came in part from the realization that they were working in a place that was special, and not just in a provincial sense. The more they studied and worked in them, the more they realized that the prairies and savannas and wetlands in and around Chicago were not just pleasant parks and lovely places for people to exercise in and enjoy, but valuable natural areas unique in the world. They supported a number of extremely rare species, such as the lakeside daisy and leafy prairie clover, each known from only a handful of dolomite prairies; the *Eryngium* root-borer moth, which survived on only a few regional prairie remnants; and the Hines emerald dragonfly, which could be found at only a few sites along the Des Plaines River southwest of Chicago and in Door County, Wisconsin. More significantly, in the minds of many, the preserves supported entire ecological communities—oak savannas, tallgrass prairies, unique wetlands—that had grown extremely rare in the entire Midwest as the region was almost entirely converted to farmland in the nineteenth and twentieth centuries.

The preserves' richness was due partly to nature and partly to people. The Chicago area started out with a rich diversity of natural settings. "We are in the middle of a convergence of major biomes here," a local conservation biologist, Tim Sullivan, once pointed out to me. "It's the eastern extent of the tallgrass prairie, the southern edge of the north woods, the western extent of the eastern deciduous forest. It's also overlaid with a very complex geological history, especially recent glaciation, that has left a diversity of soil types, topography, and wetlands. That's led to a high diversity of native species, especially plants. Fifteen hundred native plant species occur in the Greater Chicago region. That's an enormous number." The human communities of the Chicago region, through geological and ecological happenstance, simply happened to overlie a highly complex mosaic of forest, prairie, and savanna. Thanks to the foresight of the founders of the forest preserve system, the plants and animals of prairie and savanna and wetland and forest retained at least some habitat in the Greater Chicago area; elsewhere, almost their entire habitat had been plowed under for corn or soybeans. There were more than two hundred thousand acres of public land in the region that were protected from development. Chicago, people were coming to realize, didn't just have pleasant open spaces; it had, to use the new buzzword, an abundance of *biodiversity*.

The practical result of this growing realization was the establishment in the mid-1990s of a regional coalition of conservation groups and land management agencies that would coordinate restoration and conservation work across six counties in Illinois, one in Wisconsin, and two in Indiana—the entire Chicago metropolitan area. The group referred to itself as the Chicago Region Biodiversity Council, but the members realized that such a name would drop with a thud in the public discourse. Something catchier was needed. Packard, who was working for The Nature Conservancy, suggested "Chicago Wilderness." It was memorable and evocative, and it stuck. The name hinted at the richness to be found in the heart of a huge metropolitan area. It was practically a form of linguistic jujitsu, implying as it did that people—at least in the upper Midwest—could and even should be a part of wilderness. It underlined that Chicago had wild places precisely because it was Chicago, and hence home to millions of people rather than to fields of corn and soybeans. The name was, said Laurel Ross, an ecologist who was Packard's colleague at The Nature Conservancy at the time and who now works for the Field Museum, "powerful. It's fun. It's funny. It's controversial. It pushes buttons. I've never known a person who forgot it."

The formation of Chicago Wilderness was announced to the public at a party at the Field Museum in April 1996. The coalition would, it was announced, coordinate dozens of restoration-related projects in the region, thanks in large part to hundreds of thousands of dollars in funding from the U.S. Forest Service—although that agency, ironically enough, refused to accept the name. "Wilderness" means something very specific within the Forest Service. It refers to lands protected under the 1964 Wilderness Act, which called for the preservation of areas "untrammeled by man." The Chicago metropolitan area, with its eight and a half million people, was about a pole away from that definition. Instead, the Forest Service sent its money to the Chicago Region Biodiversity Council, and if everyone in Chicago referred to that organization by a different name, well, the planners in Washington didn't really need to trouble themselves about that.

Chicago Wilderness has since become a broadly recognized movement that has been emulated by conservation planners in several other cities and regions. It spawned its own magazine and has arguably done a great deal to increase Chicagoans' appreciation of the nature in their midst. That's true not only of members of the general public, but of specialists too. Said Mark Leach, a botanist and expert on oak savannas who works at the University of Wisconsin Arboretum, "The exciting thing Chicago Wilderness did was get people from the Field Museum and the universities who were used to doing research after flying off somewhere, and getting them to pay attention to their own backyards."

The coalition's formation, said Ross, both crystallized and catalyzed restoration work throughout the region. "Every county and park district now has a full-time volunteer coordinator," she told me. "They have a staff and crew for large-scale restoration. There are grants available for all sorts of projects. They can tackle jobs like earth moving or the encroachment of weedy species on five hundred acres that you can't hope to tackle with four hours and twenty people with loppers. So we have both approaches. We have projects where the volunteers who know and love every inch of a few acres and who know exactly where the prairie gentian seeds are can exercise their tender loving care. And we have large projects where an agency can go in with earth-moving equipment and lots of staff and get a lot of work done quickly."

With Chicago Wilderness, it seemed, restoration had become as much an urban institution as the Chicago Bulls or the Art Institute. It was a powerful force. Thousands of volunteers, backed by the powers and dollars of numerous federal, state, county, and municipal agencies, were working to preserve and restore rare natural environments. It seemed a perfect mix of old and new, a blending of the cutting-edge sciences of conservation and restoration with the time-tested vision of city dwellers finding revitalization in a huge city's greenbelt. Chicago Wilderness, it seemed, represented the ultimate waxing of the city's long-standing motto: "Urbs in Horto"—city in a garden.

꒾꒔

NO ONE AT THE GRAND APRIL 1996 opening party knew it, but a restoration volunteer who lived in the Old Edgebrook neighborhood on the northwest side of Chicago—in the heart of the forest preserves along the North Branch—had that month given a copy of a new book to a neighbor as a seventieth birthday present. The book was *Miracle Under the Oaks*, an upbeat account by *New York Times* science writer William Stevens of the North Branch restoration work. The man celebrating his birthday, Jim Quinn, had raised ten children in Old Edgebrook with his wife, Mary Lou. Herself curious about the restoration work, it didn't take Mary Lou long to read the book.

Old Edgebrook is a beautiful, bucolic neighborhood tucked away from the busy streets around it. It originated as a private subdivision, with a private golf course, in the 1890s, and the property that caught the Quinns' eyes back in 1958 held one of the original houses. Its primary appeal was the five upstairs bedrooms that could house their growing family, but they soon fell in love with the unbuilt areas that practically surrounded the house. Across the street was the golf course, which by then had become part of the county forest preserve system; adjacent to it were the woods along the river.

"I'm sitting here right now looking out at the snow in the woods," she told me when I called her one January day. "It's beautiful. It's totally beautiful. When we moved in we discovered it was a total paradise. It was more than I expected. We did all kinds of wondrous things here. On a snowy day most of the kids in the city are inside watching television. Mine were outside sledding or skiing or ice-skating on the river. It was paradise. So when they started cutting trees in the woods and setting fires they had me to contend with."

Mary Lou Quinn was busy raising ten kids and getting a real estate license, but still she found time to become a neighborhood activist. During the family's first few summers in Old Edgebrook neighbors would come around collecting donations for a community fund established to buy DDT. The neighborhood, well wooded and well watered, had mosquitoes in spades. She contributed, but only until she read Rachel Carson's 1962 book *Silent Spring*, which laid out in graphic detail the damage pesticides such as DDT were doing. She became a vociferous opponent of spraying by not only the neighborhood group, but also by the city and the county. Quinn cultivated political connections. In Cook County the president of the county board also oversees the management of the forest preserves, and she lobbied four of those presidents, over decades, to prevent the Forest Preserve District from fencing off the golf course. She helped campaign to stop factories upstream on the North Branch from dumping their waste in the river. She was appalled to hear rumors that forest preserve workers were cutting down walnut and cherry trees on the golf course and selling the valuable lumber.

Quinn liked to take her kids on bike rides through the forest preserves, following the same winding trail Steve Packard had, and in the 1980s she began to notice changes taking place there. Trees were dying or simply gone. "As we neared these places just north of us that used to be shady and lovely—it was a tragedy," she told me, "but we didn't know what was causing it." In their place were open, sunny patches of what she viewed as weeds. Quinn is not a naturalist. She didn't know quite what the trees were or why they were dying. She wondered if it might be Dutch elm disease.

It was not until she read *Miracle Under the Oaks*, she said, that she learned what was happening. The dying of the trees wasn't natural, it was deliberate. She was shocked. "They were killing the trees in the forest preserves to make a prairie!" she said.

She was, as it turned out, not alone in thinking it incongruous to kill trees in designated forest preserves. On May 12 of that same spring, a Sunday, the front page of the *Chicago Sun-Times* read: "Half Million Trees May Face the Ax: DuPage Clears Forest Land to Create Prairies." The article was about a new plan in the county just west of Cook, where the Forest Preserve District was planning a multimillion dollar effort aimed at restoring seven thousand acres of overgrown tallgrass prairie and oak savanna. Some local citizens formed a group called Alliance to Let Nature Take Its Course to oppose this large-scale cutting of trees.

Suddenly the Chicago Wilderness celebration in April seemed as though it might have been a bit premature. Perhaps Chicago was not quite as friendly to the idea of restoring nature as its advocates had hoped.

The columnist who wrote the story, Ray Coffey, was on the paper's editorial board, and over the next two years he wrote a cavalcade of similar articles. The headlines of some offer a practical shorthand for the questions he raised about much of the regional restoration work: "Forest Dist. 'Partners' Have Shady History." "Restorationists Gnaw at Forest Picnic Area." "'Restorationists' Talk a Chic, but Vague, Game." "Forest Preserves' 'Controlled Fires' Raise Concerns." "Forest Preserve District Is Picking Our Poison."

Quinn was thrilled to learn that she was not alone. She had been reading *Miracle Under the Oaks* as closely as a deconstructionist literary critic. "That book turned into my bible to stop this," she told me. In particular, she homed in on one passage that described some of the early restoration work in the North Branch preserves:

> So the North Branchers trod lightly, even sneakily. Their brush-clearing and planting operations took place as discreetly as possible. As a screen for their activities, they would leave a wall of brush in place along the bicycle trail or roadside. Behind the screen, they would cut a little brush here, and little there, and make their plantings. Eventually the plantings would flower and everything would look stable and lovely, and only then would they cut away the roadside screens. After the vegetation had grown up, they removed unwanted invading trees (always called brush by the North Branchers, to deflect criticism) by girdling them below the vegetation line, where the killing cut could not be seen. Then the tree would slowly fade away and no one would notice. The whole idea was to keep people from becoming upset about destroying "nature" so that nature could actually be restored.

Packard and other restorationists, for their part, deny that outright deception ever took place and point to the extensive outreach work the restoration volunteers did. "We were in a recruiting mode all the time," Packard said. "Anyone who had any interest, we'd talk to them about it. Many neighbors started getting involved. We never ran into anyone who thought it was a bad idea in the first ten years or so. . . . We started in '77, and the antagonism was in '96, nineteen years later. I don't think we'd heard much of anyone who didn't think anything was a good idea."

"We always had signs," said Karen Rodriguez, who was furious when the accusations of the antirestoration advocates hit the papers. "We always had brochures. Anyone who looked the least bit serious, we stopped and told them what we were doing. We went out of our way."

But Mary Lou Quinn considered herself informed, and armed. She formed a group called Trees for Life to advocate for an end to cutting trees in the preserves. She formed alliances with animal-rights advocates who were still upset about the killing of deer in the preserves in Cook and other counties, and together they fed Coffey a steady diet of inflammatory material. They got the ear of some of the county commissioners and of the county board president, John Stroger.

In September Stroger abruptly declared a moratorium on all restoration activities in Cook County forest preserves. It was a shock. No more workdays, no more hauling rare plants and ecosystems back from the edge! Jane Balaban was, she said, "bowled over." Almost twenty years of work, she thought, and it could come to a halt because of a few ill-informed complaints?

"There were people who had spent ten or fifteen years of their lives thinking that they were doing a good thing, and not wanting to be involved in politics," said Karen Rodriguez, "and to have their work completely stop, when they knew that stopping it for any length of time would reverse it, was heartbreaking."

When he announced the moratorium Stroger also called for a series of public hearings on the issue, which took place that October. They were lively, occupying three days and more than sixteen hours of testimony. At one long hearing, in the suburb of Skokie, more than 130 speakers testified. "What I see in the restorationists are deeply religious people," said one suburban resident, Sheldon Altman. "They suffer from God complexes. In the case of the restorationists at The Nature Conservancy, they're writing their own version of Genesis. And I can't tell if they're trying to reinvent the story of the Great Flood or Sodom and Gomorrah as they work on altering the environment."

Paul Gobster, a social scientist with the Forest Service who works in the Chicago area, listened closely to the testimonies. He heard a few speakers say that they were opposed to the idea of restoration in general, but only a few. "Even the most vocal critics would get up and say, 'We're all for restoration,'" he said, "but then they would start railing at the specifics." Some of those complaints were about particular restoration tactics—girdling trees, using herbicide—but more had to do with how decisions were made. Why, critics wondered, was the Forest Preserve District ceding its authority to a group of unelected and uncredentialed volunteers? What gave Steve Packard and his colleagues the right to decide what the preserves should look like? Shouldn't people who lived next to preserves have some say in their management?

The great majority of the speakers, though, came out in favor of the restoration work. They included ecologists and other scientists from virtually all the heavy-hitting regional organizations and agencies engaged in conservation work. They outlined all the reasons why restoration needed to happen for the sake of rare plants and animals. They talked about biodiversity and extinction and globally rare ecosystems. They presented a great deal of evidence. But they failed to entirely persuade the county board members.

"You would have globally respected people from the Forest Service, from the various agencies, the land managers, you name it, the whole scientific and land-management community, as well as many of the public, testifying in support of restoration, I think nine to one," said Jane Balaban, "and then you'd have someone like Mary Lou Quinn, who knows nothing, who gets up and rants about, 'Well, my dog died because he went in the forest preserve and it's because they put out herbicides.' And they're the same as far as the commissioners are concerned. Just unbelievable."

The hearings ended without clear resolution. The board continued to debate the issues for months, as did articles and letters to the editor in the papers and callers on talk radio shows. The delay was draining for the restoration volunteers, the criticism dispiriting. They saw themselves losing ground in the preserves, where they saw so much work to do. Finally, in February of 1997, the board approved new guidelines for restoration work that allowed volunteers to continue most of their work but severely reined in the relative autonomy they'd enjoyed. The moratorium was lifted countywide for all except the four restoration sites nearest the Edgebrook neighborhood. Girdling of trees was banned except under special circumstances. Forest Preserve District employees were to supervise restoration workdays. Detailed burn plans were to be put into place before prescribed fires could take place. And so on. The freedom of going out into the preserves and working practically without supervision was gone. So was the old sense that the work was something that—*of course!*—anyone could get behind, *the* logical goal for natural areas. A few drifted away. For those who remained, the public controversy and its aftermath served as a growing pain and learning experience. To affect the management of thousands of acres in a densely populated urban area, they realized, required more than ecological smarts and dedication to hard work. It required a commitment to the very sorts of processes that many volunteers were trying to escape on the weekend workdays.

"I look back and see how naïve we were," said Jane Balaban. "A lot of us were and many still are only interested in actually doing the work. That's what is gratifying and rewarding about doing this, is watching the health return. When this all erupted, we thought, 'Well, what you have to do is you have to go in front of the board and explain, and then they'll understand and it will be OK.' Boy, I certainly was so wrong, and so naïve! And I think one of the lessons that came out of that is, although we did not want to be political, we couldn't afford not to become political, at least to some degree. And so some of us are now involved in the politics much more than we want to be. But it's absolutely necessary."

Part of the problem that led to the controversy was linguistic. The primary architect of the work under fire, after all, was the North Branch *Prairie* Project. Many of those who testified against the restoration work were not versed in the nuance of prairie and savanna and woodland and forest. "When it began in the seventies the whole thing started with a focus on prairie, and yet it's evolved into something much bigger," said John Balaban. "As we stepped into the woodlands to continue the work out of the prairies into the woodlands, people had that thing still stuck together. So when we said, 'We're going to work in this area,' they said, 'Well, they're going to cut down all those trees and make prairie!'" Not until after 1996 was the group's name officially changed to the generic North Branch Project.

The new, more bureaucratic workday structure wasn't always workable. There weren't enough forest preserve employees to oversee workdays, and in 2002 the requirement that at least one employee be present was relaxed. In its place was what was known as the Master Steward Program: volunteers underwent training in ecology, herbicide use, fire management, and other aspects of restoration, and those certified were qualified to oversee workdays themselves.

The county board did not come out of the controversy unscathed either. The media attention cast a largely unfavorable light on its management of the forest preserves. It became glaringly apparent that the Forest Preserve District had not, unlike those in neighboring counties, hired professional ecologists to oversee restoration work (adjacent counties such as DuPage and Lake did go through their own controversies about restoration work, but they were settled much more quickly than in Cook County). Observers also raised questions about the low priority given by the county board to oversight of the preserves. In 2001 it came to light that the Forest Preserve District was suffering from a $20 million deficit—one factor that led to the defeat of five incumbent county commissioners in the fall of 2002. Of the new commissioners elected, four had made a point of emphasizing their support for restoration work.

Today the Cook County Board still oversees the forest preserves, but it does so in dedicated meetings separated from the rest of their county business. The Forest Preserve District has hired a number of trained employees to oversee and implement restoration and conservation work. They have, almost everyone agrees, a lot of work to do. In 2002 two grassroots groups, Friends of the Forest Preserves and Friends of the Parks, commissioned a study of the county's natural areas and classified 68 percent of them as being in poor shape.

Steve Packard is acutely aware of those statistics: "Even at Somme Prairie Grove—it's a beautiful place; I spend a lot of time there; a lot of people go there; they talk to me about how nice it is—you take random samples, and most of them show land in total junk. No one goes to those places, and so it's easy to forget about them." "Total junk," to Packard, means buckthorn thickets and patches of invasive, nonnative garlic mustard, a far cry from the stately oaks of the restored savanna patches. It means exactly the sort of places I explored as a kid, the sort of brushy forest patches that seemed to me at the time to be representative samples of the nature of northeastern Illinois. Perhaps Mary Lou Quinn was upset to see such places disappearing near her neighborhood, but there still are a lot of them left.

Chicago Wilderness has continued to grow. By 2005 it comprised more than 175 member organizations, which included everything from huge federal agencies such as the Forest Service and the Fish and Wildlife Service to tiny local grassroots groups and nature centers. It was sponsoring a huge number of restoration and conservation projects. It was sending volunteers out into the field to monitor birds, frogs, butterflies, and rare plants. It was compiling maps and conservation plans for the region. At Fermilab, Bob Betz was gratified to see the grounds crews he'd trained harvesting seeds and setting fires and spouting Latin binomials as readily as botanists. Chicago Wilderness had become, as its founders had intended, a tool that helped focus the energies of the entire metropolitan area.

Paul Gobster went on to compile *Restoring Nature*, a book that examines the Cook County restoration controversy in detail. When I talked to him a few years after its publication, he had just returned from studying public disputes over restoring natural areas within San Francisco. Some ecology advocates wanted to close off public access to some of those places to allow rare plants to grow, and dog walkers were upset. "Looks like I've carved out a niche for myself," he joked. In urban areas, it seems, the social complexities of conducting restoration projects echo or even surpass their ecological complexity.

"When restoration projects are situated within an urban area, people use different frames around what they're focused on," Gobster said. "Restoration advocates tend to be focused on the natural components of the site, but neighbors might have a more general view of the same spot. They might have much more utilitarian uses for the same place. As a planner you need to take those into account. Restorationists tend to be strong on science, but making the places people-friendly is often an afterthought. If you've got public lands you really can't adopt that strategy. You can make places beautiful as well as ecologically valuable.

"It's easy for restorationists to privilege the ecology, to say, 'If we lose these species we're not going to be able to get them back.' But how do you weigh? Does ecological value trump other values? If it does, we need to convince other people that it's true, but we can't ram it down their throats."

֍

MARY LOU QUINN WAS SADDENED to see the moratorium end, even though it continued in the preserves nearest her neighborhood. "It wasn't the right thing to take the moratorium off," she said. "My own heart of hearts is I wish we could get back to the 1960s and 1970s when it was shady and beautiful. It didn't get bad until man decided it was going to be a prairie. I'd like to have the prairies be nurtured back to what they should be. Plant some trees. Our area, with all these cars, we need trees." She didn't particularly care that the trees and shrubs screening Edgebrook from the surrounding streets were not native. "We're all immigrants from somewhere," she said.

Quinn admits that she is not a scientist or a naturalist. But her opinions on nature are strong and heartfelt. She simply sees trees as providing urban benefits, such as shade and protection from noise and pollution, that prairies cannot, and she considers the biodiversity that restoration advocates point to—the hundreds of species of plants that make up a prairie or savanna—a canard. "A prairie is *one* thing—a prairie," she said. "It's *trees* that purify the air."

Her opinion of the restoration work was also colored, though, by her long experience of observing the county government and bureaucracy at work and by what sounded to me like a visceral dislike of Steve Packard. She accused him of profiting off the restoration work by using it to gain lucrative government grants. At a community meeting in Edgebrook she heard Packard say that the North Branch workers had "listened to nature" and had realized that some of the forest preserve land should be oak savanna rather than prairie. She took that, and the change of the group's name from North Branch Prairie Project to North Branch Project, as an admission that they had no idea what they were doing and were looking to deceive both politicians and public.

"Listening to nature!" she retorted. "I'm surprised he's not talking to God. How does he know about oak savannas? He had no, none, zero scientific background—no more than I have. Nature doesn't need Steve Packard. It can take care of itself."

Quinn didn't go as far as to say restoration was a bad idea, period, but after two hours of conversation I could discern no way in which she would allow it in her vicinity. Her beliefs about what nature should look like were far too strong for that. She saw the condition of the forest preserves as she came to know them, decades ago, as a natural condition, and would do anything to return to that. She was unconcerned about the fates of rare plants or animals or ecosystems. And, given her long experience of Cook County's government, she was ready to see corruption everywhere.

"Restoration, even the name, is a good thing," she told me. "It's a lovely word, like *religion*. But people are abusing it. You take a good thing like biodiversity or religion and somebody's going to corrupt it. Most of the restorationists are good people. They just got misdirected. People became cultlike in saving the prairie. They take the volunteers and brainwash them. It's the fad of the times. You make mistakes. We're all human. You build dams and find out later you shouldn't have. When I was a child doctors were advertising Camel cigarettes."

Trees for Life was fairly quiescent at the time of our talk, although Quinn was staying involved in politics and going to county board meetings to argue her position. Another advocacy group called Natural Forest Advocates (motto: "The best way to be a friend of the forest is to leave it alone") remained active and was busy searching out objections to prairie and savanna restoration in the scientific literature. It sent out a bimonthly newsletter that went to all the county board members. And Quinn intended to continue working to maintain the restoration moratorium in her neighborhood.

⁂

ONE OF THE SITES WHERE most restoration work continued to be banned under the county moratorium was Bunker Hill Prairie. Volunteers were allowed to collect seeds and remove a few species of nonnative forbs, such as garlic mustard, but they weren't allowed to cut trees or conduct prescribed burns. Quinn was convinced that restorationists were continuing to sneak in to cut trees and apply herbicide. But the Balabans, the site's stewards, could hardly stand to go there anymore. "We go there three times a year," said John. "We get a garlic mustard workday, a white sweet clover workday, and a seed-collecting workday, and other than that we pretty much stay away because it's too painful. But yeah, there's a tremendous growth of young green ash, et cetera, that needs to come out of there, and I've certainly led that workday in my head fifty times already. I know exactly what I'm going to do as soon as they say, 'OK, you can go out and cut the green ash.'"

What they saw on those three annual workdays, though, was enough to further persuade them that they were doing the right thing. The prairie plants continued to thrive even if thickets of green ash were coming in. To their eyes the place was visibly much healthier than it had been. "Painful as the moratorium is," said John, "we were able to see that Bunker Hill, which had maybe fifteen years of work at the time of the moratorium, has held on pretty nicely for the last eight years, despite our not being able to do what needs to be done there."

୬ଡ଼

WHAT IMPRESSED WILLIAM R. JORDAN III, back when he attended his first Chicago-area restoration workday in 1990, was the way the tools were lined up. Jordan, who lived in Madison and worked at the University of Wisconsin Arboretum, was attending the second annual conference of the Society for Ecological Restoration. He had helped form the organization and had conceived and founded the journal *Restoration & Management Notes*, but this workday was his first hands-on experience in Chicago.

"Whoever was running this field trip had arranged the tools in this military sort of way," Jordan said. "They weren't just thrown there on the side of the road; it was as if we were a rifle company.

There was a protocol, a *style* to it. I remember getting off that bus and thinking: 'That's classy.'"

Steve Packard was on that field trip, and Jordan was impressed by him too, especially by the way Packard recited the scientific names of the seeds the restorationists were scattering. "At the conclusion of our two hours out there someone said, 'My God, it's even in Latin, like the old mass,'" Jordan remembered. "We all got a kick out of it."

In the world of ecological restoration Jordan has been the philosopher who looks for deeper meanings rather than the activist who gets his hands dirty and pitches the politicians and mobilizes the grassroots, a sort of Simone de Beauvoir to Packard's Betty Friedan. During his almost twenty years of editing *Restoration & Management Notes* (which was eventually renamed *Ecological Restoration*) he became known as a thinker who always strove to widen the discourse about restoration beyond the mechanics of raising native grasses and eradicating invasive plants and reengineering wetlands, writing editorials, for example, that explored the ethics and implications of restoring such places as a Nazi concentration camp. In his 2003 book *The Sunflower Forest: Ecological Restoration and the New Communion with Nature*, he broadened his explorations of restoration in a manner that takes the word *communion* quite literally.

Jordan, who moved to the Chicago area to teach after leaving the arboretum in 1999, looks like a philosopher. He is tall and balding, appears a bit owlish, and likes to wear berets. He liberally salts his conversation with references to Robert Frost and Henry David Thoreau and Joseph Campbell.

Jordan trained as a botanist and began working as a public outreach specialist at the arboretum in 1977. His job essentially involved marketing the place, and as he pondered its selling points he came to dwell upon the story of Aldo Leopold's restoration work. Leopold had died in 1948, but his posthumously published book *A Sand County Almanac* had by the early 1970s become one of the bibles of the environmental movement. Dog-eared paperback copies bounced around in a million backpacks. Leopold's pioneering work in restoration, however, had been largely forgotten, and his version of preservation, at least as it was rendered by most others, meant leaving wild places and wild animals alone. They needed to be protected, advocates argued, because human beings could only harm them. North America had started out as a pristine continent that people, through their manifold acts, had only degraded. Nature was perfect; people were fatally flawed. The possibility of restoration, many believed, was a dangerous one in a political and social climate of exploitation; if damaged places could be repaired, it was all right to damage them in the first place. It was the same argument, Jordan realized, that Robert Elliot was making in Australia; it was why Bob Betz was so sensitive about people knowing that he scattered seeds in prairie fragments.

"Arguing for the protection of natural places," Jordan wrote in *The Sunflower Forest*, "environmentalists stressed their vulnerability, often insisting that they were not only 'fragile' or susceptible to human or 'outside' influences but were actually 'irreplaceable.' Though understandable, this sort of rhetoric and the thinking it represented had devastating implications for conservation. It implied that conservation was a one-way street, essentially nothing more than a delaying action, that might slow the inevitable decline of natural landscapes toward eventual extinction but can never reverse it. It also conveyed the idea, often expressed quite explicitly by environmentalists, that the influence of human beings on natural landscapes is invariably negative and destructive: though we may take from such a landscape, we can never give anything back to it. Such thinking . . . is deeply pessimistic."

It was so pessimistic, Jordan told me, that he had come to think of the neglect of restoration as "one of the great defining mistakes of twentieth-century environmentalism." Perhaps the sort of work Leopold and John Curtis had pioneered at the arboretum offered another alternative. Perhaps it pointed the way toward a human engagement with the natural world—which Jordan likes to call "classic landscapes"—that could be gratifying and humble, actively engaged and respectful. Restoration, he concluded in *The Sunflower Forest*, "combined the best elements of two forms of environmentalism —the conservationist's willingness to participate in the ecology of a natural landscape, and the environmentalist's insistence on the inherent value of that landscape, independent of its value to humans—into a single act that linked engagement with total respect."

"I don't want to say there's a set of ideas here that can save the world," he told me, "but they might point in that direction. Restoration is *the* paradigm for the conservation of classic landscapes—it's the best chance we've got."

As Jordan worked at the arboretum and edited his journal, he quickly came to see through the work of Packard and many other people that restoration did have a profoundly important role to play in preserving endangered species and ecosystems. The evidence of that filled issue upon issue. What came to interest him even more, though, was the effect restoration had on its practitioners. That restoration was so necessary in so many places seemed to him evidence that the relationship of people to their surroundings was, if not entirely broken, then at least badly bent. In *A Sand County Almanac* he read Leopold's admonition that people ought to fix that relationship by becoming "plain members" of the land community, rather than conquerors of it. What, Jordan wondered, did that mean? What might it mean to be a *member* of an ecological community, particularly at a time in history when Americans in their isolating cars and gated subdivisions and Internet reliance seemed to be fleeing from any ideals of actual human community? He was reading a lot of history and anthropology, and saw how people had traded in sitting around the campfire and telling stories for going to the theater where professionals acted out those stories, and then they'd traded in watching live actors for watching a movie, and then they'd traded in even the mild camaraderie of the cinema for the isolation of sitting at home in front of the TV.

"By most interpretations," he told me, "community means less and less every generation. Both the experience and the institutions of community have arguably weakened progressively since—I don't know when. Community, like all the higher, transcendent values, is not just easy. It is emotionally demanding and involves confronting some aspects of our relationship with others that we don't find particularly or at all attractive." It is because Americans these days don't really have to navigate the treacherous shoals of community to survive that many choose to do without it—although many Americans also seem to spend a lot of time decrying its lack.

It is through ritual, Jordan came to realize, that people since time immemorial have dealt with many of the most problematic aspects of relationships with other people: both through the big, self-conscious rituals such as weddings, funerals, and hazings, and also through countless ingrained acts such as shaking hands, a small and often practically mindless ritual that began as a very pragmatic means of showing a potential adversary that you weren't carrying a dagger. If ritual remains so important in acting out tricky relationships with other people, he thought, shouldn't it also be important in negotiating our equally complex and difficult dealings with nature?

Jordan's musings on these questions came up at about the same time as restoration advocates in Illinois and many other places began taking on truly thorny issues, such as the question of whether it was justified for people to shoot deer so as to preserve plants and ecosystems people wanted to see. As a botanist, he could see that restoration was vital to those plants and ecosystems, but perhaps it was equally important to its practitioners. Perhaps restoration might serve as a sort of ritual for, as he put it, "negotiating our relationship with the landscape, for getting close to it, understanding it, for developing a sense of caring about it." Because restoration relied so heavily on ethically fraught tactics such as cutting down unwanted trees or shooting what were perceived as excess numbers of deer, perhaps it was uniquely suited to helping people come to terms with the entire constellation of human impacts on the landscape.

One day Jordan stumbled across an essay in *Harper's* by a literary critic named Frederick Turner, who raised many of the same questions. The two struck up a correspondence. Turner argued that such "transcendent values" as beauty and community could be achieved only through a full appreciation of the difficulties inherent in them—such as the troubled human interactions in any community or the violence inherent in living on this planet and surviving through the eating of animals or even plants. Only by fully embracing those difficult experiences, Turner argued, could people truly feel the benefits of community. "See, at some point we have to connect with the rest of nature," was how he put it in an essay about abortion, "and it always involves death."

Turner's parents were anthropologists who had worked in Africa and had written extensively about performance studies. *Performance,*

Jordan realized, was simply another term for ritual. Here was the link he'd been looking for. People in many traditional societies perform explicit rituals that help bond communities internally and with their surroundings, often by embracing horrific violence. On the Great Plains, Native American Sun Dancers pierced their own breasts and ripped their flesh in painstaking obeisance to religious laws. Some of the people of what was to become Papua New Guinea ritually killed and ate young men and young women of their own tribe during puberty ceremonies. In the Judeo-Christian tradition Abraham was prepared to sacrifice his own son to God and only at the last moment found his hand stayed so that he could kill a sheep instead. The Christian mass very specifically involves flesh and blood. Such rituals, Jordan said, are "technologies of the imagination. They've made killing into an occasion for celebration so that it *isn't* murder." The various forms of sacrifice, rather, are a way to symbolize the debt that human beings owe to nature. "What we get from the rest of nature is *everything*, and we know in our heart of hearts we can't pay that back," he said. To believers it is vital to acknowledge and underline that debt through symbolism and ritual, and for that reason such rituals are not pro forma. They are truly needed to maintain order in the world.

What restorationists were doing, Jordan came to believe, was exactly the same: using controlled violence to maintain order in the ecological world, and doing so by working in a creative zone of tension between what we tend to term the "human" and the "natural." As an example, Jordan pointed to the work of an artist, Barbara Westfall, who lives near Madison. When staff members at the arboretum were restoring a patch of prairie by girdling the bark of some aspen trees, she turned the killing into a work of art. Rather than allowing the stripped trees to fade into the background, she highlighted the girdling by removing yet more bark and by applying paint to emphasize the wounds, thereby turning, as Jordan wrote in his book, "what might have been a routine, clinical procedure into a sacrificial act and an occasion for the creation of beauty." It was the ritual, in this case expressed as a work of visual art, that sacralized the taking of life. Even the North Branch tradition of the bagel break halfway through a work outing, he has suggested, could be viewed as a ritual that lends meaning to the acts of restoration that take place around it.

Although he does not specifically discuss it in *The Sunflower Forest*, Jordan casts new light on the Chicago restoration controversy. Those who have most vehemently opposed the killing of deer, and the cutting of buckthorn, act out the belief that nature can only be demeaned by contact with people. They retain that traditional American view of nature as perfect and people as perpetually flawed. The only responsible thing to do, then, is to leave nature alone so that some place, however small, might retain some grace.

"Those of us who are inside the forest, the woodlands, the prairies, looking at them, we can see what's happening inside, or what's not happening inside that should be happening inside, and as a result restoration makes sense to us," John Balaban commented. "The fact that people actually have to get involved and do something is clear and obvious. But the people who are opposed to our work, I would describe as people who stand outside the forest preserve. They stand outside and it's green. And as far as they're concerned it's green and it's the way God made it, and it's the way it ought to be, and everything is happening according to plan."

The restorationists, on the other hand, accept the blood and dirt of human agency. They believe that people can and should actively improve the forest preserves. When they look at a preserve filled with buckthorn they see not an area dominated by natural processes, but a place profoundly impacted by people. It makes no sense to say that it should be left alone now when it has not been left alone—not in an ecological sense—since the great booming metropolis was established in the early nineteenth century. To them ecological realities are much more pressing than abstract notions about what nature could or should be. If zigzag goldenrods evolved in and relied on particular conditions in the Chicago region, people have an ethical responsibility to see to their persistence into the future.

One way to limn the debate is to examine what that tricky word *nature* means to those who support or oppose restoration work. Antirestoration advocates see the forest preserves in their current condition as nature, and they say that allowing excessive human manipulation in them is unnatural. If zigzag goldenrods can't tolerate today's conditions, that's simple natural selection. Restorationists, on the other hand, argue that each natural area has a particular trajectory, or some condition it should be in, and that any well-intended human effort to put the place back on that trajectory in the face of human neglect is part of nature, too. Bill Jordan likes to talk about how restoration entails what he calls "a studied disregard of human interests." Instead of deciding that a particular patch of land should be, say, a picnic ground, or a softball field, or something else that has a particular use to people, restorationists decide that it should be what it should be— which is to say, what it would be if nature were truly in charge. It's really a broader definition of nature, even if restoration advocates may have a narrower view than some of their opponents about the ecological conditions that should prevail in any particular place.

It's a tricky definition, too, because of course every place that's restored still *does* have a human use as open space or a savanna grove to look at or a place to go birding—which can give restoration opponents a lot of ammunition. A restored savanna at Somme Prairie Grove still has some human intention embedded in it; it is what it is only in part because of the natural processes of seed dispersal and plant growth and the foraging of animals, and in part because of conscious decisions made by human beings expressing a certain set of values. It's tricky terrain, philosophically speaking, and that is precisely what Bill Jordan sees as so promising. Because a landscape being restored is neither purely nature nor purely manufactured, it is rich, suggestive, even liminal, full of symbols and freighted with meaning.

In Jordan's view the killing that is the weightiest element of restoration work exactly parallels the killing that people need to do to survive, and people need to acknowledge it, own up to it, to avoid being demeaned by it. That sort of mindful sacrifice is part of the broad definition of nature that fuels restoration work. I cannot say that such mindfulness accompanies the cutting of every buckthorn stem along the North Branch, but after talking to Jordan I thought it entirely possible that many of the volunteers may have been consciously or subconsciously building rituals—particular ways of conducting prescribed fires, or even the lowly bagel break—that may with time take on a broader meaning.

Packard hinted at the difficulty and the promise of establishing a new relationship with nature in an article he wrote on the savanna controversy: "I, too, was once drawn to nature as a place without people—a place for contemplation. I suppose I still am. But species and whole communities are vanishing and will disappear utterly if we just watch. The new view of nature is admittedly poorer in romantic purity and mystic detachment. Yet it's richer in participation. The values that are being lost are not entirely to be mourned; to a considerable degree they were a product of our alienation from nature. The restoration ethic allows us once again to belong in nature. Throughout most of our species' history, we were a part of nature. Our challenge now is to rediscover that role and play it well."

Perhaps the careful laying out of the tools on that long-ago field trip, in other words, was an incipient ritual in the same way that Packard's early projects had been, according to Betz, "incipient prairies."

⁂

MY CONVERSATION WITH JORDAN was very much on my mind during the 2003 winter solstice celebration held by the North Branch volunteers at Somme Woods. It was a blustery late afternoon, but not particularly cold by Chicago standards. The tops of the trees heaved in the wind; the sky was clear. It was just a little before sunset when a crowd of about 150 people gathered in the parking lot, herded together by the wailing of a bagpipe.

When the bagpiper stopped Steve Packard began to speak. He was wearing jeans, a dirt-colored Carhartt jacket, and his ancient floppy hat.

"This is part of one of human culture's oldest traditions," he said, "the turning point of the year. It's a time for many people to think about the previous year, how it went, and the next, how it will go."

Packard thanked the forest preserve staff and its commissioners, specifically mentioning Greg Goslyn, one of the new commissioners elected on the reform ticket a year earlier, who was in the crowd. I'd just spoken to Goslyn about his support for restoration work. "The Forest Preserve District," he'd said, "has come to realize that there are hundreds of people out there who will eagerly do for free the sort of work that we hate to do on weekends at home."

Packard spoke of how rose-breasted grosbeaks and eastern wood-pewees and blue-spotted salamanders had showed up in the Somme Woods' restored areas. He invoked Aldo Leopold's call that people ought to be members of the land community.

"Some of these oak trees the buffalo walked under, and the Native Americans, and the mountain lions, but they didn't walk through these dense thickets of buckthorn. They *did* walk under the oaks.

"One of the things we're doing here is consecrating the new year. Today we have the opportunity to commune with nature, to commune with our neighbors, and to watch a very powerful force," he concluded.

The bagpiper led the procession into the woods on a muddy trail. We wound our way under the big oaks of the old ghost savanna and through buckthorn thickets and past a small marsh crusted with a thin sheet of ice. After a couple of hundred yards we entered a large clearing—a new clearing, with countless small stumps still sticking up among the downed leaves—around a huge brush pile the size of a delivery truck. It was meticulously built. Its insides were made of buckthorn trunks and branches, many of which still bore brown and shriveled leaves. Above and around that was a framework of ash logs.

John McMartin, who again was wearing his flame-resistant coveralls, recruited five crews of helpers to place wads of newspaper just inside the pile. Then he handed out boxes of matches.

Packard spoke again. "We encourage everyone to mark your own personal solstice," he said, "by gathering up a stick and investing it with something you want to remember from the last year, something you want to forget from the last year, something you want to hope for in the coming year, and when you put it on the bonfire it can be spiritually important to you."

Sticks began to fly, some overshooting the pile and landing among the onlookers on the far side. There were a lot of enthusiastic kids in the crowd. McMartin's helpers lit their matches and then the newspaper. The flames sputtered and, on the upwind side, quickly went out.

Jane Balaban had talked, back at her house, about how natural fires could not spread on the northern Illinois prairies anymore. The landscape was too cut up by roads and buildings to allow that. "We always have to be the lightning to start the fire," she'd said. It seemed a good summation of the responsibility of people to the natural world, at least in a heavily built-up place like Cook County. If the smooth phlox and the burr oaks and the zigzag goldenrods are going to persist here, it will be because people like the Balabans are taking an awful lot of trouble to see that they do.

On the downwind side the flames caught and spread. They spread through the newsprint and into the most delicate of the buckthorn twigs and licked up into and through the huge mass of branches until they shot up twenty feet and the trunks of trees on the far side came to waver in the heat haze. They rose high while the crowd of people assembled there from city and suburbs threw sticks in with hopes and wishes known only to themselves and helped themselves to hot chocolate and apple cider and cookies. They danced hypnotically as the sky grew dark and skeins of ring-billed gulls flew high overhead, heading out to the lake for the night from some inland landfill or parking lot Dumpster. They lofted even higher after Steve Packard walked around them recruiting help in tossing heavy eight-foot-long green ash trunks onto the fire, while John McMartin used a long, straight pole to tidily heave the unburned ends of logs into its glowing heart. The flames kept on long after dark and remained a beacon in the woods as many of us walked back to the parking lot. A Northbrook police officer had just driven up. He was looking at the orange glow through the buckthorn thicket and asking, "Is someone going to turn it off?"

No, no one was. Whatever one might prefer to call it—a prairie fire or savanna fire or solstice bonfire—it was clear that it was one fire that was not going to be extinguished any time soon.

3

The Entrepreneurs

*B*ack when Bill Jordan used to run into Steve Packard at the annual conferences of the Society for Ecological Restoration, he and Packard used to compare the history of restoration to that of early aviation. After the Wright Brothers' first flight, there were plenty of attempts to bring aviation to the public. But despite the barnstormers and the Red Baron, it remained a novelty well into the 1920s, a diversion with, seemingly, little practical application. Not until Charles Lindbergh flew across the Atlantic in 1927 was the public truly gripped by flight. Only after Lindbergh, Jordan and Packard realized, did aviation reveal its practical potential; only after his flight did the public broadly come to understand and dream about the possibilities for airmail, for commercial travel, for military use. Only after his flight did Americans imagine themselves flying for some reason other than entertainment. Only after Lindbergh, in other words, did flight enter the mainstream imagination and the mainstream economy.

As they worked in the 1980s and 1990s, Jordan and Packard mused on how restoration needed to find its "Lindbergh's flight." They knew that even the thousands of volunteers working in the Chicago forest preserves were a tiny percentage of the region's population and that their passion was still viewed as a fringe interest, a hobby. It was telling that the restoration outings there took place mainly on weekends rather than weekdays. If restoration were ever to truly enter the mainstream of American society, it would have to enter the American consciousness the way the dream of the skies had in the late 1920s. It would have to have a personal effect on the lives of many people, not just on nature enthusiasts. It would have to both capture the imagination and play a role in many people's pocket-books. It would have to become something that people did during the workweek. It would have to become a part of the country's economic life.

When he came to visit the southwestern ponderosa pine forest, Jordan believed that he had perhaps found his Lindbergh's flight. The forest is, he learned, both extensive enough and messed up enough to require restoration on a vast scale. The acreage is impressive. Ponderosa pine forests cover millions of acres in Arizona and New Mexico alone, along with millions more in other parts of the arid intermountain west. You could, were you so inclined, saddle up a horse on the outskirts of Silver City, in southwest New Mexico, and ride a long arc to the north and then the northwest, past the small city of Flagstaff, Arizona, and not lose sight of a ponderosa pine tree for a distance of more than 250 miles. You'd circumambulate mountains whose cold heights are crowned with dark spruce and fir trees. Your horse would pick its way through the grassy hummocks of broad meadows and through dry boulder fields of limestone or basalt where soils are thin and small clumps of cactus grow among the rocks and the trees remain skinny and gnarled. On this long ride you'd get a feel for how ponderosa pines spring so reliably from the arid highlands of the West, clinging to life even on steep cliffs and rocky mesa edges, thriving on the extensive upland flats, growing to their greatest height and girth in the deep fertile soils of moist valley bottoms. On hot summer days you would smell the sun-warmed odors of vanilla and butterscotch emanating from the pine's thick bark, which peels off in wafer-thin flakes shaped like jigsaw puzzle pieces and whose color on mature trees approaches that of the flesh of sweet potatoes. You'd see the boles of the biggest trees, a yard and more in diameter, reaching more than a hundred feet high like great architectural columns and bearing at perpendicular angles stout branches tipped with long needles, and you'd hear the wind in those broad canopies, and you'd be thankful for the delicate shade they cast.

All that is true. You might see the largest ponderosa pine forest in the world as one of the West's great tracts of open land. If you happen to come to it, as most people do, from more populated places, you might see the ponderosa pine forest as the naturalist and Chicago native Donald Culross Peattie described it in 1953. It was, he wrote, "the West's prime 'Vacation-land.' . . . Its dry and spacious groves invite you to camp among them. Its shade is never too thin and never too dense. Its great boles and boughs frame many of the grandest views, of snow-capped cones, Indian-faced cliffs, nostalgic mesas, and all that brings the world to the West's wide door." You might see it, in other words, as a cardinal example of wild land—land where the natural has not yet been displaced by the human; where nature proceeds as it always has, independent of human desires; and where a wide range of people find recreation, meaning, beauty, and sublimity precisely because of that independence.

On the other hand, you might, as increasing numbers of people have, see this vast forest as an enormous problem. You might see it as a series of enormous fires waiting to happen. You might believe that after absorbing the influences of modern civilization for well over a century the forest bears little resemblance to what it once was and has about as much to do with the West's original nature as does some tacky interstate tourist trap hawking mass-produced arrowheads and Indian blankets made in Mexico. You might, cold-eyed, assess how much people have altered the forest's long-standing ecological cycles. You might, in short, view this vast tract as a headache, a liability. All that is true too.

When Jordan first visited Flagstaff in 2002 the latter view was markedly more prevalent than the former. The southwestern ponderosa pine forest was, as he put it later, a place where "when you mess it up, nature comes back to bite you in the butt." Jordan visited Flagstaff in July of that year to teach a seminar at Northern Arizona University, only a few weeks after Arizona had had its butt bitten in a big way. On June 18, the largest wildfire in the state's recorded history broke out just below the Mogollon Rim in the east-central part of the state. It was a ferocious inferno that created its own weather and sent embers hurtling miles away. More than thirty thousand people, including the entire population of the small city of Show Low, were evacuated from their homes in its path. Flagstaff, where I live, is more than eighty miles away, yet in the evening we could see the fire's immense smoke plume from there, churning and billowing like an enormous gray-white thunderhead. It was visible from space. The TV newscasts and front pages carried apocalyptic photos of burned-out cars, towering walls of fire, and weeping families. When it was finally out, more than two weeks after it began, the Rodeo-Chediski fire had charred more than 465,000 acres of forest and more than four hundred buildings. Property damage costs totaled at least $120 million. The money spent in attempting to suppress the fire and in the rehabilitation of the lands burned exceeded $175 million.

The fire broke out during conditions of high wind and severe drought, but its fast spread and its severity were widely ascribed to the unhealthy condition of the area's forests, especially the ponderosa pine forests that made up the bulk of the acreage burned. It was in line with a clear trend of ever larger and more serious fires in the Southwest's ponderosa pine forests:

> 1977: La Mesa fire, New Mexico: 15,270 acres
> 1990: Dude fire, Arizona: 24,174 acres, 6 lives lost,
>   63 buildings burned
> 1996: Dome fire, New Mexico: 16,516 acres
> 1996: Horseshoe/Hochderffer fires, Arizona: 24,765 acres
> 2000: Cerro Grande fire, New Mexico: 47,650 acres,
>   235 buildings burned
> 2002: Rodeo-Chediski fire, Arizona: 467,066 acres,
>   426 buildings burned

These infernos are not the kind of fires Steve Packard's friends set in the Chicago forest preserves. They are fires that roar across the land, unstoppable, consuming all the needles of hundred-foot trees in seconds, practically immune, even, to all the technologically sophisticated blandishments of the modern firefighter: quick-response systems, air tankers, fire retardants. They are souvenirs of more than a century of ecological meddling and especially of the exclusion of the ground fires that once maintained ponderosa pine forests. Lacking small fires, the forests grow thick with small trees and become ripe for conflagration. They present a grave danger to the human communities set within the forest and, to Jordan's mind, a signal opportunity for restoration. Here, he thought, the scale of the ecological problems instigated by modern humans had grown so great that restoration had the opportunity to become the dominant paradigm defining how people manage the land; in fact, it would have to.

After Jordan went back home he added his thinking about the northern Arizona landscape to his book's final manuscript. "The situation here is dramatically different from that on the prairies," he wrote, "which die if deprived of fire, but die quietly, leaving the system intact as the prairie is gradually replaced by another kind of community, often oak forest. The ponderosa pine forests, in contrast, die violently, striking back viciously and self-destructively, demonstrating in a way the prairies do not the worst consequences of human neglect. If it was on the prairies that conservation learned the value of restoration, it may be in the ponderosa pine forests that the rest of us will at last learn its importance."

Maybe so. If that is to happen, though, it is likely that some reconciliation will have to take place between those two primary ways of viewing this extensive forest: as a monumental wilderness and as a monumental problem. If that is going to happen, it will require that the region's people find, through the practice of restoration, a tricky middle way between two deep-rutted and comfortable paths.

⁂

To understand where those two paths diverged, imagine standing at the Flagstaff railroad station in the early autumn of 1896. Two men get off an eastbound train. The younger man is dark and luxuriantly mustachioed and carries himself like the patrician he is. The other is almost twice his companion's age, and his long beard and deep-set eyes give him the appearance of an Old Testament prophet. They look around at Flagstaff, a rowdy wood-frame logging town, home to just over a thousand people and about a dozen saloons and bawdy houses, that is struggling to dust itself off and practice respectability. Then they get into a stagecoach for a bouncy ride to the Grand Canyon.

The two men are Gifford Pinchot and John Muir, and they are on a tour of the West's forests as part of a federal commission charged with evaluating the region's newly designated timber reserves, the forerunners of the national forests. This stop is near the end of their tour. They have already witnessed huge clear-cuts in the Big Horn Mountains of Wyoming, in Idaho, and in northern California, and horribly overgrazed forests around Crater Lake in Oregon. The scenes are in keeping with the era. Wanton exploitation of the nation's natural resources is the order of the day. The bison have been practically extinguished from the Great Plains, the tallgrasses are rapidly giving way to John Deere's plow, the great pineries of the Great Lakes states are almost gone, passenger pigeons and Carolina parakeets and Labrador ducks have pretty much vanished from the skies. The United States may be immeasurably larger than a constricted island like Bermuda where such resources as cahows and sea turtles were practically wiped out within decades, but in the 1890s it is rapidly becoming apparent that it, too, is subject to certain ecological limits. Some officials in Washington are taking note of that. A few are rousing themselves to do something.

The two men have seen ample evidence of overuse in the terrain around Flagstaff. In some places the train traversed virgin stands of yellow-barked ponderosa pines that stood tall and open and sunny. Elsewhere, though, it passed enormous stump fields from which loggers had taken every single sizable pine for miles. Northern Arizona, one visitor to Flagstaff had written a few years earlier, held "an almost limitless supply of the finest timber in the world." That *almost* was a surprising note of caution in the heady, boosterish hayride of the 1880s. The town's first sawmill opened in 1882, and soon thereafter cleared lumber from its abundant old-growth pines was going into mine timbers in Tombstone, Arizona, and Durango, Mexico; into boxes that brought California oranges, apples, and cantaloupes to market; and into pilings, sidewalk planks, shingles, shipping boxes, and construction materials from Colorado to New York. Thanks to the very railroad the two men rode—which itself spread its weight on ties of ponderosa pine—Arizona pines had become part of the world economy as soon as the loggers arrived.

Pinchot and Muir also didn't pass an acre that hadn't been grazed. Only a few years earlier, in the late 1880s, some two hundred thousand sheep had grazed the San Francisco Peaks area near Flagstaff.

They were an efficient way to convert the accumulated sunshine and snowfall of virgin terrain into dollars, and for a few years the owners of those first flocks made profits of something like 50 percent per year. Cattle roamed everywhere too. But like the high-tech boom of the 1990s, it couldn't last. When severe drought hit the Southwest in the early 1890s, enormous numbers of cattle and sheep died. In some parts of the Arizona Territory, it was said, you could stand on a cattle carcass and throw a rock onto another, and onto another from that one, for mile after denuded mile. Many of the first ranchers went bust. The grasses were grazed to their roots and the dirt lay bared, and it gullied and eroded and ran off red and ocher in the summer rains.

Such was the landscape Muir and Pinchot traversed to get to Flagstaff, a land that only twenty years after the first American settlers arrived was showing extensive signs of wear. They must have talked about how to respond to those signs. It's too bad no one in their train compartment was recording their conversation, because for those interested in how people have lived in western North America for the last century the encounter of these two men in this landscape is about as compelling as it would have been for a student of religious schisms to witness Martin Luther and the pope palling around. Here were the two chief proponents of the two chief land-management strategies that have been at play in the West ever since. Pinchot, soon to be designated head of the new federal Forest Service, was the era's principal proponent of scientific forestry, of the wise use of resources; Muir, soon to become head of the Sierra Club, was the greatest advocate of valuing wilderness for its own sake, of leaving nature untrammeled.

In the absence of a written record we have to imagine how they interpreted what they saw. They would have agreed that the enormous clear-cuts and the excesses of grazing were an outrage. After that they would have differed. Muir believed that nature could not be improved upon, whereas Pinchot claimed that cool, rational management would allow humans to control the unruly forces of nature. Muir must have seen in the stately groves of ponderosa pine natural cathedrals. Pinchot must have visualized how a new regime of scientific management would allow opportunities for new towns, new sawmills, and new productive employment in this frontier region, while leaving plenty of trees and grasses to provide for future needs. Today the landscape they traversed shows how both men had their political victories. The South Rim of the Grand Canyon, where the two men camped together after their bumpy stagecoach ride, is part of Grand Canyon National Park, which was established by President Theodore Roosevelt—who also, while president, went camping with Muir—and has been managed primarily according to his dictum that "you cannot improve upon it. The ages have been at work on it, and man can only mar it." Most of the forestland around Flagstaff, on the other hand, came under the aegis of the U.S. Forest Service, the agency Pinchot came to manage when Roosevelt created it in the winter of 1905. This forestland was one of the new federal reserves Roosevelt was talking about when he said, on the occasion of his first State of the Union address, "The preservation of our forests is an imperative business necessity." On them Pinchot's ideas held sway. The federal forests were to be shielded from the worst of the overexploitation rampant on private lands, but they were emphatically to be *used*. They would provide timber and livestock grazing opportunities and jobs for a growing nation; they would become part of the country's booming economy.

The rangers of the new national forests were, as Roosevelt liked to say, "men with the bark on." With millions of acres to oversee in the new national forest system—and Roosevelt's swagger behind them—they were feeling their oats. They could handle guns and horses and weeks living out of the saddle on beans and coffee and wild game. Given that, they could certainly handle the forest. They'd seen the great pineries of the North Woods and the Southeast cut almost entirely over. They had science on their side. They knew there was a better way. If the loggers could be kept from cutting all the trees for miles, if fires could be stopped, and if some pines could be made to regenerate, the loggers would be protected from their own greed. They might have a crop in perpetuity, rather than another bust once all the big trees were cut.

After the Forest Service was formed, Pinchot must have thought back on the trip he and Muir shared to northern Arizona. He bore in mind a recollection of seeing those huge clear-cuts in which not a single seedling pine was sprouting to replace the cut trees. In 1908 he directed a young forester, Gus Pearson, to set up the nation's first federal forest experiment station in an unlogged tract just northwest of Flagstaff. Fort Valley lay right along the old stagecoach route. Pearson's task was to figure out how to grow young ponderosa pines there and create a new crop.

The trouble, as Pearson found within the first few years of his long residency at the Fort Valley Experimental Forest, was that most years the northern Arizona climate is too dry to allow seedling ponderosas to survive. He experimented incessantly with nursery trees to learn if there was something he could do to help them along. It took ten years before he learned that he didn't have to bother. Ponderosas weren't in trouble, he learned; they simply have an episodic mode of reproduction. In 1918 the mature pines around his station bore a splendid cone crop. The following year brought exceptional rains in May—nearly three and a half inches—followed by a wet and warm summer. It was, Pearson wrote, "just like heaven." By 1920 the seedlings were as "thick as the hair on a dog's back." That summer was dry. Many seedlings died, but more than enough survived. Grazing was still heavy, and the seedlings didn't have much competition from grasses. The cutover lands bristled with new pines. It was, Pearson thought, a great success story—tomorrow's crop sprouting and growing tall in the clear sunshine left by yesterday's. It was exactly how the science of forestry—and the national forest system—was supposed to work. Pinchot must have been proud.

He and Muir, by the way, ended up sharing a fine night together on the South Rim on that 1896 trip. They stood on their heads on the rim to better appreciate the canyon's wild colors. Muir dissuaded Pinchot from killing a tarantula, pleading that it had as much right to live as the two men did. When night fell the two spontaneously decided to camp on the rim like schoolboys on holiday. They shared a campfire and a long evening of stories and didn't return to the hotel until dawn.

In coming years, though, their friendship was to grow strained beyond the breaking point. Muir continued his uncompromising defense of wilderness; Pinchot became a pragmatic politician who continued to advocate for efficient, scientific management of wild

places. Early in the twentieth century the two parted ways entirely
over the proposed damming of Hetch Hetchy Valley in the Sierra
Nevada. Their dispute wound its way down through the twentieth
century in this form: Pinchot's adherents came to believe so thoroughly in human efficiency and effectiveness that they saw no reason
to leave an acre alone; Muir's were so sure of the benefits of untrammeled nature that they had serious reservations about integrating a
single wild forest acre into the human economy. The schism, to be
sure, was often exaggerated for political purposes: environmental
writers especially, the historian Char Miller has pointed out, came to
"set the devil Pinchot against the angel Muir." Their trip to northern
Arizona, though, helped catalyze two movements that, in broad
strokes, came to dance around one another like opposing magnetic
fields. Muir: The meaningful products of the forest are its spiritual
gifts, and we can gain those only by leaving it alone, by doing no
more than walking through it in rapture. Pinchot: We can manage
the forest better, more efficiently, than nature can, and there is no
harm if we can turn a tidy profit in the process. It was, after all, the
American century, an era of riches and progress for all.

⚬

IN NO WAY DID PINCHOT and the other men of the new Forest Service show that they could outdo nature more than in their treatment
of fire. Fire was endemic in the forests of western North America.
Early American explorers witnessed it all the time. John Wesley Powell, first explorer of the Colorado River in Grand Canyon and the
chief of many federal survey parties, recorded how the smoke of the
Southwest often made it difficult to draw accurate maps: "A haze of
gloom envelops the mountain land and conceals from the eye every
distant feature," he wrote in 1890.

   It was a matter of simple chemistry. The West is above all a region
where moisture comes irregularly, and at some point virtually all its
forests become dry enough to burn readily, even explosively. In the rain
forests of the Pacific Northwest a drought lengthy enough to kindle big
fires might come along only every few hundred years, but in much of
the interior West dry seasons come frequently, practically yearly. When
Euro-American settlers arrived there, fires were common.

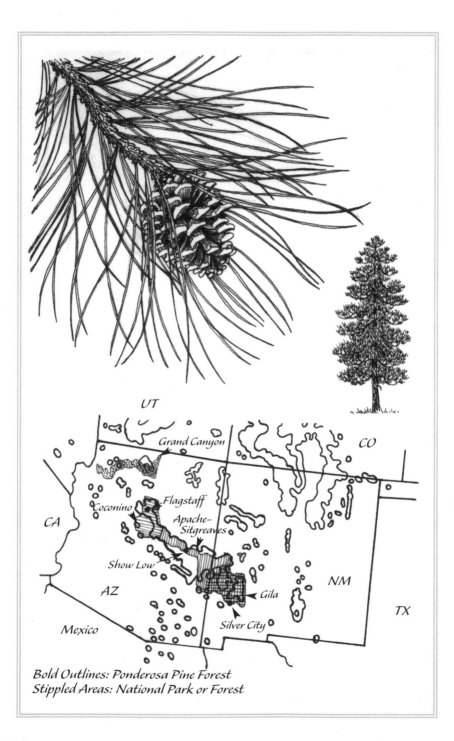

UT

Grand Canyon

CO

Coconino

Flagstaff

CA

Apache–
Sitgreaves

Show Low

NM

AZ

Gila

TX

Mexico

Silver City

Bold Outlines: Ponderosa Pine Forest
Stippled Areas: National Park or Forest

To Pinchot and his colleagues, any fire roaming the landscape was wanton. Fire, they were convinced, was a force that ought to serve humans. It was to be safely corralled in the manifold new engines of the industrial era. It was there to extend man's reach, to be useful. Forest fire was anything but. Forest fires burned up wood that ought to become timber and grasses best suited to filling a cow's belly. Forest fires were capricious and dangerous and wasteful—and inefficient. Science and careful management, the new forest managers were sure, could get rid of them. The suppression of wildfires became an imperative business necessity. "To-day we understand," Pinchot wrote, "that forest fires are wholly within the control of men."

In the earliest days of the logging industry in the Southwest, before the Forest Service, fire truly was a serious problem: it was driven largely by all the slash the loggers left drying on the ground and by the sparks cast off by logging railways. As early as 1885 loggers and sawmill crews were sent to fight forest fires near Flagstaff. By the early twentieth century the Forest Service organized firefighters as one of its first tasks. The first set of instructions to the first supervisor of the San Francisco Mountains Forest Reserve, headquartered in Flagstaff, were issued in 1898, and they included the directive: "It is of the first importance to protect the forests from fire."

By then, though, it was hardly necessary. The sheep and cattle had eaten all the fuel. Lightning still struck and campfires still escaped, but there was no ready way for fire to spread along the ground anymore. That was fine, thought the foresters; that way, all those new pines that sprouted during moist years, such as 1919, could thrive. And they did, for a while, even as attitudes about fire became more and more entrenched. In 1910 massive fires in the northern Rockies killed eighty-five people and raged over millions of acres. The Forest Service adopted a blanket strategy of fighting all fires. In 1935 it instituted the "10 A.M. policy," which mandated that all fires were to be under control the morning following their discovery. Fire was the enemy, and it was to be given no quarter. To waver in opposing it would mean the defeat of all the well-meaning ideals of scientific forestry. It would mean that people were not in charge. On the Great Plains the dust was billowing off the dry fields and blowing east until it darkened the skies over the nation's capital. That wasn't part of the manifest destiny script. Americans needed to be reassured that they were, somewhere, in charge of nature. They were about to find an icon for exactly that.

In the Depression summer of 1933 guests at the Circle Flying W dude ranch, just northwest of Flagstaff, found a deer fawn, seemingly abandoned, in the nearby woods. It was adopted at the ranch and became a great favorite of the guests, part of the charm of their rural vacation in the sticks. That summer the guests happened to include Hazel Sewell, who was Walt Disney's sister-in-law, and Joe Grant, who was one of the Disney studio's chief artists. You ought to sketch this cute fawn, Sewell suggested to Grant. He did, and before too long the idea for *Bambi* was working its way through the new Disney studio. When the movie came out in 1942, it crystallized the modern attitude toward forest fire. Fire was the fault of people; it devastated the forest; it robbed helpless animals of their homes. The forest was better off without it, and so were we.

By then a Japanese submarine had shelled California, and officials were growing concerned about incendiary balloons floating east from Japan, one of which actually did manage to kill six civilians when it exploded in Oregon in 1945. Getting materiel to the troops overseas required a lot of wood. Preserving the western forests from fire came to be seen as not just a business but as a patriotic necessity. The War Advertising Council and the Forest Service inaugurated a campaign that encouraged Americans to prevent forest fires. Bambi was the first spokesanimal, but he was available only briefly. There were contractual issues; he had business elsewhere. A new symbol was needed. An artist sketched a bear, and Smokey was born. He was named for an actual New York City firefighter. When in 1950 a real live bear cub was rescued from a fire in New Mexico, a real Smokey was born. He went to live at the National Zoo, in Washington, D.C., and proved so popular that he became the only American celebrity with his own zip code. Thanks to him, preventing fire became a byword for civic responsibility.

It was, experts agreed, one of the most effective advertising campaigns ever. Also effective was the drive to fight fire once it began, at least in the Southwest. For decades after the institution of the Forest Service policies, even into and after World War II, most of the fires in the Southwest were small and easily controlled by a handful of men with mattocks and shovels. All throughout the ponderosa pine belt those seedlings that Gus Pearson had seen in the spring of 1919 grew into saplings and then into the middling-sized trees that foresters call poles. Photosynthesis did its work and the pines reached up and out. Kept safe from fires, the trees grew, the timber industry thrived, cattle grazed, and everything seemed fine. The Forest Service was working out just as had been planned. Pinchot's cool rationality and efficiency had, it appeared, prevailed.

🐦

GIFFORD PINCHOT would have recognized much of the education in forestry that Wally Covington underwent at Yale in the early 1970s: the calculations of board feet, the smooth curves illustrating idealized distributions of tree sizes, the measurements of forage production. By the time Covington arrived in Flagstaff as a newly minted forestry professor in 1975, however, a few cracks were beginning to show in the smooth façade of the Forest Service's management of its forests. That year Covington began working on a research project with two Forest Service scientists, Steve Sackett and Jack Dieterich. In 1974 the two researchers had set up a project to study the role of fire in the ecology of ponderosa pines at a place called Chimney Spring, on the Fort Valley Experimental Forest. If a few people in the 1960s and 1970s were becoming aware that prairies needed fire to survive, others were developing the same idea about ponderosa pine forests. Sackett and Dieterich wanted to test that notion.

It wasn't hard to do. Mature ponderosa pines, with their thick bark, don't burn up readily when grass fires pass by. A fire can, however, burn away a bit of the bark and leave a small patch of heartwood exposed, which chars readily when other fires pass. By cutting out small samples of the growth rings of old trees, whether living or dead, Sackett and Dieterich were able to count how many times the trees had been charred. Further, ponderosa pines are exquisitely attuned to the often-pronounced annual fluctuations in the western climate, making it relatively easy to date them by variations in the width of their annual growth rings. A trained viewer can read those

rings as readily as an accountant reads a spreadsheet and can deci-
pher exactly when the charring fires occurred.

Sackett and Dieterich took samples from the trunks of seven large
pines, all well over two hundred years old. What they found was
striking. The trunks contained a great many char marks, which
recurred as often as every two or three years. One venerable tree that
had sprouted in 1440 displayed a continuous record of thirty-one
fires over more than three centuries. Over the entire study plot the
average interval between fires was only about five years. Clearly, fire
had been common in this forest during the lifetime of these trees.

Two facts about the fire scars stood out. First, all the trees were of
a fair size and age when they were first burned. One of the trees had
already put on sixty-three growth rings before it showed its first char
mark; the rest were all older. Trees much younger than that probably
didn't survive the fires. They burned up rather than charring. Their
bark was too thin or their flammable needles too close to the ground.
These old trees, it appeared, were rare survivors that had been lucky
to avoid the frequent fires during their youth. Sackett and Dieterich
had read historic accounts, like that of the explorer E. F. Beale who,
about forty years before Muir and Pinchot, passed through the area
that would become Flagstaff. "We came to a glorious forest of lofty
pines," he reported back to Congress, "every foot being covered with
the finest grass, and beautiful broad grassy vales extending in every
direction. The forest was perfectly open and unencumbered with
brush wood, so that the travelling was excellent." It was fire, then,
that had maintained these dry forests in this open condition. It raced
through the drying bunchgrasses and fallen pine needles every few
years, likely during the dry months of early summer, burning hot and
clear—but it *raced*, devouring those light fuels and most saplings, yet
for the most part passing by too quickly to do much harm to the
older pines.

A second fact was far more striking as the researchers considered the
implications of their findings: not a single tree showed any sign of a fire
after 1876, the very year in which the first American settlers arrived in
the Flagstaff area. That was well before federal fire suppression policies
began, a lifetime before Bambi, and long before Smokey was telling
campers to put their fires out, but it was right about the time that the

first sheep and cattle arrived. Once the livestock ate away the grasses, the researchers realized, ground fires had no way to spread. That allowed many more of the seedlings that so pleased Pearson to grow. If the old trees recorded the absence of fire in their rings, the young ones echoed it in their very presence. The Chimney Spring plots were visibly much denser than they apparently had been before the fire cycle had been disrupted; where each had had perhaps a few dozen large trees per acre before 1876, they were now thick with hundreds of midsized pines per acre that in the old days would have burned up.

Some foresters had already written of the importance of fire in thinning ponderosa pine forests. A handful had experimented, mainly on tribal land beyond the ken of the Forest Service, with using prescribed fire to burn up the accumulating grasses and needles that over time came to represent a more and more severe fire hazard. Sackett and Dieterich wanted to assess whether such fires might also thin out some of the smaller pines. They knew that fires before 1876 had likely burned in the hot, dry months of early summer, but that was too dangerous now. With Covington's aid, they ignited their fires in the fall instead, burning different plots at different time intervals: yearly, biennially, and at other intervals up to ten years.

The fires did kill a few of the smaller pines, but not as many as the researchers hoped: the "doghair" thickets described by Pearson decades earlier were generally too shady and too moist to burn easily. A great many of the big old-growth pines, though, were killed by the fires. It wasn't because their needles caught fire—they were too far off the ground—and it wasn't because their bark wasn't thick enough. They died because the thick layer of downed needles around their trunks smoldered for many hours or even days. The pines' upper roots were simply baked, as was some of the vulnerable cambium tissue within their lower trunks.

"The experiment was set up to show that getting fire back into the forest would take care of the problem," Covington said. "But within three years we found that wasn't the case at all. Where we wanted fire to burn hot it burned too cool, and where we wanted it to burn cool it burned too hot." Clearly, using prescribed fire by itself in dense, modern stands would not re-create the conditions Sackett and Dieterich found recorded in the Chimney Spring fire scars.

Covington was hooked. He believed that he had seen not only the past, but also the future of the region's forests. In much of the upland Southwest, he realized, conditions were as good as at Chimney Spring for figuring out what the forest had been. American settlers had arrived late here, after the arrival of photography and the passage of well-trained naturalists who did a pretty good job of keeping notes. And in the region's arid climate the fire scars and other evidence of historic forest structure remain well preserved for many decades. Why not use that evidence as a model for management? he wondered. It would be easy to thin today's dense forest back to something like what it had been before fires had been disrupted; then it should be possible to use fire to maintain it without the negative tree-killing side effects he'd seen at Chimney Spring. It would be much healthier than leaving the forests to their own devices, which to him meant burning them up in destructive fires. He was ready to pick up the drip torch—but, before that, the chainsaw.

"The kind of thinning you could gain from prescribed burning was inadequate to return to natural conditions, and it killed the old-growth trees," he told me. "I think I became the most strident advocate of both thinning and burning to restore more natural conditions. I was looked at kind of awry for that. People didn't really see the implications of it, except for maybe saying, 'We need to thin a bit.' Nobody was saying, 'We need to get it back to what it was.'"

Since the 1970s numerous other projects have buttressed the findings of the Chimney Spring work. Tree-ring studies in many parts of the Southwest have shown that ponderosa pine forests did indeed burn frequently: near Flagstaff, every four to seventeen years; at Mount Logan north of the Grand Canyon, every five or six years; at Five Pine Canyon in the foothills of the San Juan Mountains of southwestern Colorado, every five to twenty-three years; in the Jemez Mountains of northern New Mexico, every six years or so. Forests that had once supported dozens of trees per acre now supported hundreds, but many of those were in poor health, overstressed by competition for water and nutrients, susceptible to drought and to attacks by bark beetles. Throughout the Southwest massive changes had taken place in a forest that, to the untutored eye, looked pristine. As in the Chicago forest preserves, only because change occurred so slowly, over the course of an entire human lifetime or more, did no one really notice.

People did notice changes in how fire behaved, however. From the establishment of the Forest Service through the 1960s large fires were hardly an issue in southwestern forests, although they did occur in other parts of the West. But Covington's arrival in Arizona coincided with an alarming increase in the size and frequency of fires on the national forests of the Southwest. In 1971 and 1974, for the first time since records had been kept, more than two hundred thousand acres burned; in 1979 the total exceeded three hundred thousand acres. Not all were in ponderosa pine forests, but many were.

Yet more alarming than any numbers were the effects of these fires. Rather than moving along the ground, swiftly devouring the grasses and dried needles, the fires traveled through the tree crowns, killing young and old pines alike. Where they burned particularly hot they devoured the soil's organic matter and left behind bare mineral dirt that ran off after summer thunderstorms. In many cases they were so disruptive and destructive that it was rank weeds rather than pines and native grasses that grew back after the fire had passed. They represented a real danger to people too. They were fires that got attention. What they showed was that suppressing small fires, year after year, was no way to get rid of fire entirely. It was, though, a good way to concentrate fire, to ensure that it would be severe and costly once the inevitable spark landed. Years of treating fire as the enemy had, in the end, resulted in infernos that were just like the horrific fire in *Bambi*.

༠ঌ

IN THE EARLY 1990S COVINGTON was finally able to put into practice his ideas of getting back to the way things were. He and several Northern Arizona University colleagues received a National Science Foundation grant to conduct a restoration experiment on a ten-acre tract of forest at the Gus Pearson Natural Area, immediately adjacent to the Fort Valley Experimental Forest headquarters buildings. The tract had never been logged, and massive yellow pines still towered high. But from ground level it was hard to see them, so inundated were they in a sea of small, young pines. By mapping the location and determining the age of every single living tree, dead tree, and downed log, Covington and his colleagues were able to assess the extent of the changes that had taken place. Where only twenty to twenty-five trees per acre had stood in 1876, about 1,250 grew per acre in 1992. And it wasn't just the trees that had changed. In 1876 only about a fifth of the ground surface lay in shade at noon on a summer day; the rest was the grassy openings Beale had described two decades earlier. In 1992, thanks to all those doghair thickets, only 7 percent of the stand remained open. This part of the experimental forest had not been grazed since 1910, but it hardly mattered; in the shade of all those small trees the grasses had no way to take over their old role.

The experiment was classically simple. On two-thirds of the stand Covington's team cut down almost all the trees that had sprouted since 1876. They left standing all of what came to be called the "presettlement trees," meaning those that had been there in 1876. In an effort to replace the old trees that had died since then they also left a few younger trees standing wherever a downed log or standing dead tree showed that a pine had stood. Then they burned half of the treated tract, after first piling grass cut from a nearby meadow onto it to mimic the fuel load that would be present in an open ponderosa stand and after raking the piled dead needles away from the big trunks. The other third of the stand was left untouched as a control area.

The results were stark. The old pines in the treated area showed increased vigor and health, as measured in such attributes as the toughness of their foliage and the uptake of water and nitrogen. Even more striking was the change in the forest floor. Grasses and wild-flowers teemed in the thinned areas, both burned and unburned. Three years after the tree cutting there was about twice as much herbaceous vegetation there as in the control area.

All those effects have been carefully monitored since 1992. You don't need to be an ecologist, though, to understand the effects of the initial cutting and the subsequent prescribed fires, which have taken place every four years since 1992. The thickets in the control area are choked with small trees crowded together like riders on a Tokyo sub-way, many no bigger around than a drink coaster. Quite a few of them are bowed over, their spindly trunks unable to support their weight.

There isn't much plant life growing among the mats of downed needles on the forest floor. Beale might be able to walk through this forest, but he'd have a hard time getting his horse through.

The adjacent treated area looks entirely different. The grand yellow pines stand tall. They're venerable, with thick, platy bark and
massive, gnarled limbs. They're a few hundred years old and, with
luck, will live a few centuries more. Cutting out the abundant young
trees had the same effect on them as thinning the lettuce shoots in a
garden: with less competition for nutrients, they've thrived.

What's also noticeable about the treated area is that there are
scarcely more than two sizes of trees: the big old pines and a scattering of younger ones, many stemming from the bumper crop of 1919,
twenty and thirty feet tall. It will take those small trees a long time to
reach the stature of the yellow pines. Then again, impatient people
generally don't become foresters.

Looking at this collage of old and young trees is a bit like standing in a Chicago oak savanna and seeing the ghost of an old landscape emerging from the constrained circumstances of the present. It
looks good, even to those who will never live long enough to see the
younger trees grow to old-growth stature.

⊰%

FOR COVINGTON, the Gus Pearson experiment was a ringing
endorsement of his restoration strategy. The combination of thinning
and burning worked. The old trees survived and became healthier.
The remaining young trees, freed of most competition, could more
quickly grow to an equal stature. The grasses and wildflowers
thrived, and the risk of high-severity fire ripping through the tree
crowns was much reduced.

He could have left it there. Covington could have spent the rest of
a respectable career working among those same pines and teasing out
more and more details about how they interact with grasses, insects,
fire, and one another, but he didn't. Each time he flew out of Flagstaff
he saw the vast carpet of ponderosa pine stretching off beyond the
horizon, fine-textured with myriad small trees rather than coarse-
textured with old-growth pines and grassy meadows. He worried.
He sees that landscape as a very big problem.

"I think we're going to have more big fires," he told me a couple of years after the Rodeo-Chediski fire. "We're very vulnerable. My guess is that over the next thirty years we'll see restoration efforts stepped up as lives are lost, houses are burned, watersheds lost. There's still this human tendency to expect that probability won't catch up with us. But I think it will. It's going to take a few more calamities to really get things going."

Unlike a great many academics, Covington has not shied away from the world of policy and politics. He is a tall man who attracts attention in much the same way that a lightning rod does during a thunderstorm. He has a broad face and a down-home manner that betrays a boyhood spent in Oklahoma and Texas. He is genial and is comfortable with the sort of backslapping small talk that is far more endemic to the world of politics than that of academia. From the time he was a junior professor he made a habit of calling government offices in Washington and Phoenix. I've got some forest research I'd like to show you, he'd say, that's pertinent to this problem we're having with fire. He liked to make the calls in summer, when a typical day's temperature in Phoenix is 105 degrees. At seven thousand feet, Flagstaff is high and cool. Covington had visitors. Over the years he came to have good relations with representatives and senators and governors. The research organization he founded at Northern Arizona University, the Ecological Restoration Institute, has been very successful at garnering millions of dollars in federal funding (I was hired by the institute in 2001 to edit a book about ponderosa pine forest restoration). Covington has, a colleague said once, an "evangelical approach" to restoration. Propelled by a sense of urgency, he has been a sort of academic entrepreneur who stirs things up, who gets noticed. Someone was going to have to stir things up, he reasoned, if restoration was to hit the big time. The inevitable criticism was simply a cost of doing business.

Cutting trees was a hard sell in the Southwest in the 1990s. In late 1995 a federal judge issued an injunction stopping all logging on national forests in Arizona and New Mexico, claiming that the Forest Service had not adequately analyzed the effects of logging on the threatened Mexican spotted owl. Tempers flared. Logging firms and sawmills shut down. Although the idea that there were far too many trees in the region's pine forests was taking hold, there was no longer any practical way to remove those trees.

Things came to a head in 1996. It was a drought year in the Southwest, and the forests dried out early and stayed parched in an early summer heat wave. The fire season began early and lasted long. In Flagstaff it crackled to life on June 20, when a lightning strike ignited a fire that burned through the Hochderffer Hills just northwest of the city. A pall of smoke hung on the horizon. No one knew which way the fire, which was soon entirely out of control, would spread. Shortly another fire jumped a two-lane highway and joined the first. Together the Hochderffer and Horseshoe fires burned almost twenty-five thousand acres. Flagstaff, it became clear, had had a close call. Only luck had dictated that the fires did not begin near any of its numerous forested subdivisions. Firefighters began to repeat a dire stock phrase: it was not a matter of *if* neighborhoods would burn, it was a matter of *when*. Clearly, they said, it was time to clear out some of the heavy forest fuels around town. Cutting trees and setting prescribed fires were becoming a civic responsibility. "Stumps and smoke" became a mantra, a goal that would reflect progress.

One morning that summer Covington got a call from Bruce Babbitt, one of his political buddies. Babbitt, who'd grown up in Flagstaff, was a former governor of Arizona and, at the time, the secretary of interior under President Bill Clinton. In ten minutes I'm going on the *Today Show*, he told Covington, and I'm going to tell them that your Flagstaff model is the answer to the forest fire problems the West is having. You better be ready to answer some questions about it for the media.

There really was no such thing as a Flagstaff model, yet Covington was ready. He had his spiel all worked out. We know what we need to do, he said, and we need to do it on a large scale. If forest fires are going to occur on a scale of tens of thousands of acres, we need to be planning restoration treatments on that sort of scale too. Putting in more treatments on the scale of the Gus Pearson project might look and feel good, but wouldn't do a thing to reduce the very real fire danger and ecological degradation facing western forests.

The restoration of the southwestern ponderosa pine forests, in fact, appeared to be one of the few ecological issues that *could* be effectively explained in a morning news show sound bite. It was far simpler than messing around, over generations, with the genetics of the American chestnut. It involved none of the distinction between native and nonnative plants that was in that same year causing such controversy in the Chicago forest preserves. The places in question, after all, were still pine forests, still wildlands, and almost all the original ecological pieces were still in place. All the forests needed were a bit of thinning and a bit of fire—a bit of tweaking of ecological processes—and they'd be fine. In other places the practice of healing an entire region's ecology might be only tenuously connected to the interests of mainstream society; in the ponderosa pine forest, with its newly chastened human communities, the two were almost identical.

It all sounded so simple. With Babbitt's help, Covington got the go-ahead to experiment with the restoration of thousands of acres of remote forestland managed by the federal Bureau of Land Management around Mount Trumbull in far northern Arizona, north of the Grand Canyon. In Flagstaff a coalition of environmentalists, land management agencies, and ecologists formed a nonprofit organization eventually called the Greater Flagstaff Forest Partnership (GFFP), which set for itself the goal of restoring, annually, ten thousand acres of the fire-prone forest that girdles the city. It all had a breathtaking sweep. In one fell swoop the practice of restoration was poised to expand from small projects on tens of acres to huge ones on tens of thousands. Lindbergh's plane, it appeared, was heading on down the runway.

ॐ

BUT THERE WERE ISSUES. The flight was turbulent, the final destination uncertain. At Mount Trumbull some of the early prescribed fires that followed thinning treatments scorched and killed some of the remaining trees, even big yellow pines. Cheatgrass, a nonnative species that has invaded millions of acres in the Great Basin, moved in, displacing native grasses and wildflowers. Environmentalists, some of whom were ready to see a Trojan horse full of loggers whenever they saw a researcher with a chainsaw, thought the place looked terrible, full of stumps and churned-up soil. "In twenty years Mount Trumbull will be seen as one of the greatest debacles of the century,"

Flagstaff environmentalist Sharon Galbreath told me in 2002, citing concerns about erosion, the cutting of large trees, and effects on wildlife species. Wildlife biologists worried that Covington's cuts were too radical and left too little habitat available for such species as woodpeckers, squirrels, goshawks, warblers, and other animals that live in tree trunks or canopies. At Fort Valley, where the GFFP put in its first treatments across a few hundred acres of Forest Service land, several environmental groups formally appealed the agency's decision to go ahead with the project. "Do You Have to Destroy a Forest to Save It?" asked a full-page newspaper ad from the Southwest Forest Alliance and the Sierra Club—echoing rhetorical techniques pioneered by the latter organization's erstwhile director, David Brower, four decades earlier—that damned the Fort Valley treatments as "extreme logging." To some environmentalists, it was too cozily convenient that a forest ecologist was suddenly calling for the cutting of so many trees so soon after the shutting down of the region's timber industry.

The logged areas in Fort Valley did, in fact, look very unlike the open, lovely stand of big trees at the Gus Pearson Natural Area. Most of Fort Valley didn't have any big yellow pines left; they'd all been cut decades earlier. As a result, Covington's thinning prescription reduced dense stands of small trees to open stands of small trees that didn't look anything like the historic photos of the Flagstaff area before logging. If you looked at them in a certain light, they looked like a tree farm; in another, they looked like the germ of a future old-growth stand.

Covington was unfazed by the criticism. He reached often for the same medical analogy that Steve Packard likes to use. When a patient comes back from open-heart surgery, he was wont to say, he looks terrible for a while. That doesn't mean he didn't need the surgery; he'd be far worse off without it. Covington has a sense of conviction about the work he does. He has never liked the commonly used term "presettlement model" for his restoration plans, and I asked him once what he did want to call it so as to distinguish it from other thinning-and-burning prescriptions being developed in the Southwest.

"Ecological restoration," he said, after a moment's pause. "Putting things back the way they were, the way Steve Packard does."

Some of the most thoughtful critiques of his working plans came from a conservationist named Todd Schulke, who lives near Silver City and who had, in 1989, been one of the founders of an environmental group that eventually came to call itself the Center for Biological Diversity. The center is a hard-litigating, take-no-prisoners nonprofit organization that was one of the central players in shutting down the old-growth logging industry on the Southwest's national forests in the early 1990s. That campaign was wildly successful: in 1986 more than thirty million board feet of lumber were cut on the Gila National Forest; twelve years later, none. Schulke, though, is not a no-cut activist. "We started seeing the writing on the wall," he told me. "We'd been after the Forest Service to stop cutting old growth. But now there was this clear problem with fire. We saw that from a management standpoint the Forest Service had much stronger arguments for thinning than it ever had had for old-growth logging."

Schulke, in fact, collaborated with Covington on a couple of experimental projects in New Mexico and in northern Arizona, but came to disagree with what he saw as both Covington's tendency to cut too many trees and his fixation on matching the locations of trees in the contemporary forest to those in the historical forest. Covington is a fundamentalist: he believes that the traces of the old forest, read literally, provide the best guide for the new. "Maintaining present conditions is kind of like engineering a forest without a blueprint," he said. "What restoration provides is the blueprint. Without reference conditions it's pretty hard to say you're doing restoration. Reference conditions are, as Steve Packard says, not just a point in time, but the last, best information we have on how the ecosystem sustained itself."

Schulke agrees that it is important to know as much as possible about what a forest stand once looked like, but after some experimentation he came to disagree about the importance of leaving trees standing only where logs or stumps showed that old trees had once been. He saw how that tactic often led to the cutting of large trees that happened to stand in what once had been grassy openings and to the retention of small trees that happened to stand near relic stumps or logs. He was more interested in retaining as many big trees as possible; after all, he'd spent years opposing the logging of big trees precisely because they'd become so uncommon in the ponderosa pine forests. Cutting too many big trees could be devastating to some wildlife species, he said, and could become an economic incentive for the Forest Service.

"We really take exception to the idea that trees or clumps of trees need to be in historical locations," Schulke said. "It makes more sense to use what's there. That helps achieve a balance between protecting wildlife and reducing fire risk. It also removes the potential for pulling larger trees off for the sake of timber volume. It takes the economic drivers out of the marking system."

Such disagreements have played themselves out, time and again, even on restoration projects that practically everyone agrees are necessary. How many trees should be cut? Which ones? Coming to agreement on precisely how to thin trees, however, has not been the only obstacle to implementing restoration on a broad scale. The prescribed burns that are supposed to follow thinning have also proven a challenge in many cases. Reintroducing fire to the landscape, the historian Stephen Pyne has written, is "akin to reinstating a lost species"—meaning that it is far trickier than simply reversing the process by which it was practically extirpated in the first place. Things have changed while it's been away. Its habitat has been altered. It needs to be finessed, taken care of.

Returning fire to the forest requires stepping back from Pinchot's philosophy of rigorous management and control. Up the hill from where I live the fire department set a prescribed fire one windy spring day. Weeks later the ground was still blackened, the smell of smoke still in the air. Walking the terrain, I could read in charred trunks and scorched needles how the fire had behaved. In most places it had remained on the ground, among the grasses, but in some it sprang up, browning lower branches or even entire canopies in the thickets of small trees. It did more or less what the fire crew wanted, but, clearly, it was also unpredictable. Where it blew up might be due merely to a chance gust of wind or to a dead branch dangling toward the ground. A chainsaw is precise; wielding it, you know exactly which trees are going to live and which will die. With a drip torch, you take your chances, and the forest is sculpted in a different way. In other words, letting fire do the work—even a closely watched prescribed fire—entails a degree of letting go.

☙

SOMETIMES THE DISPUTES about how to conduct thinning treatments and when to set prescribed fires are reminiscent of the medieval arguments about angels dancing on the head of a pin: people who are in 90 percent agreement about what to do fight ferociously and righteously about the remaining 10 percent. Often that remaining 10 percent amounts to only a few trees per acre, which seem to carry a lot more ideological than ecological weight. Yet in the long run these disagreements about just which trees to cut, and difficulties in reintroducing fire to the landscape, are likely to prove only a sideshow diversion from the real challenges. Those challenges are economic. You can't restore a landscape of millions of acres with the same techniques that have proven so effective on Nonsuch Island and in the Chicago forest preserves. These days there's no Civilian Conservation Corps to work more or less for free, and volunteers certainly can't come anywhere near handling the workload. Meanwhile, the Forest Service and other land management agencies face ever more strapped budgets. Both at Mount Trumbull and around Flagstaff the ambitious plans to restore thousands of acres a year quickly slowed to what seemed a crawl—often a few hundred acres a year—due to protests from environmentalists, to the slowness of bureaucratic decision making, to difficulties in finding loggers to do the work, to funding limits. Stumps and smoke were proving challenging goals.

Of course, restoration treatments on a truly large scale could considerably reduce fire danger in the Southwest and could eventually reduce the considerable expenditures spent in attempts to corral huge fires like the Rodeo-Chediski, but such long-term investments seldom have much currency in federal budgeting. Assessments of the acreage of ponderosa pine forest that needs to be treated so as to reduce fire danger run into the millions of acres in the Southwest alone. Someone has to be paid to wield the chainsaws. In an era of shrinking federal budgets that is likely to happen only if some market can develop for the material removed from the woods—if there is some reason to wield the saws that is integral to the human use of commodities. Back when the logging of the old-growth pines was running at full steam, after all, there were good economic reasons to do so, healthy profits to be made. That's how things happen in America. And that's

why a great many people have spent a lot of time trying to figure out
how to integrate forest restoration into the modern economy.

Think of the puzzle this way. There are many things you can do
with a piece of wood, but at its core a tree is sunlight captured and
converted by photosynthesis into potential fire. You can cut the tree
and burn it in a campfire, on a hearth, in a woodstove, but in those
forms wood is a balky fuel, irregular, inefficient. It is no wonder that
wood-burning locomotives gave way long ago to coal, which in turn
gave way to diesel. Gas and oil and coal—which can readily be pul-
verized into a fine powder—are far more practical fuels for an industrial economy. They can be transported in pipelines and fed into
engines as gas or liquid. They've been standardized. Pump a gallon of
gas into your car and you know pretty much how far it'll take you,
but how long will that fresh log in the campfire last? Well, that
depends: on the species, on the wood moisture, on the way it's cut,
on the night's humidity. Wood in its raw form can feed a very small-
scale economy; it can heat your camp or your house. It can fire your
pottery, grill your steak. To feed a large, centralized economy, though,
you need fuels that flow like water.

It was to make wood move a bit more like water that a man
named Rob Davis came to Show Low in 1992. He wanted to turn
the wild variability of wood into a tame product, a commodity: a
pellet, to be precise.

Show Low is probably the only town in the United States named
after a hand in a card game; the winner, back in 1876, won the ranch
where the town would one day be located. The winning hand? The
deuce of clubs. Davis knows something about luck too, or at least
about gambling. In the 1980s he was a contractor building pumping
stations for the Central Arizona Project, the huge system of canals
that transports Colorado River water to Phoenix and Tucson. It was
good, lucrative work, but he tired of collaborating with huge engineering firms. He wanted to do something on a smaller scale. In
1992 he talked Japanese investors into backing the construction of a
mill for making wood pellets in Show Low.

To most investors the idea of going into the wood-products industry in northern Arizona in the early 1990s was roughly akin to bank-
rolling a buggy manufacturer in 1920s Detroit. The logging industry

there was in its death throes, strangled by a dearth of large trees, by competition from other states and other countries, by a flourishing of environmental regulations. The smart money was off elsewhere. The only winning way to invest in Show Low, anyone with dollars and sense could tell you, was to build more houses in the surrounding forest and sell them as cool, shaded summertime retreats to Phoenicians tired of the desert heat.

Davis disagreed and played his hand. He established his company, Forest Energy Corporation, in an open woodland on the edge of town. Thirteen years later, when I went to talk to him, it was still there, humming along and bigger than before.

Davis is wiry and loquacious, a lean coil of energy. He will be glad to tell you the following facts about turning wood into pellets. A chunk of ponderosa pine that occupies a cubic foot weighs sixteen to eighteen pounds. A cubic foot of pine pellets weighs forty-two pounds; it contains commensurately more Btus and produces more heat than the unprocessed wood. Because pellets are consistently sized and shaped, they burn far more efficiently than wood logs. A new pellet stove produces about a fifth the emissions of a new woodstove. In most communities it can be used even when "no-burn" days are declared for the protection of air quality.

When Davis began production in 1992, he primarily used sawmill waste and the shavings produced at molding, truss, or furniture manufacturers—mostly in Phoenix—as his raw materials. His business grew steadily, if not spectacularly. Pellet stoves had only begun to be sold in the United States in the 1980s. Demand was limited. It's grown since then, although not nearly as much as Davis would like. A month before I visited him he was at a trade show in Austria. Twenty makers of pellet stoves were represented there, all but a few from Austria. Austria has a population of about eight million. The United States and Canada together, with a combined population of more than 346 million, have about twenty-three pellet stove manufacturers. Wood pellets, unlike fossil fuels, are considered carbon-neutral. Burning them produces carbon dioxide, but it can be taken up again quickly by more trees, whereas it would take eons to make more oil or coal. As a result, consumers in countries that have signed on to the Kyoto global-warming treaty and that assess a tax on carbon don't pay such taxes on wood pellets. Sweden is a signatory to Kyoto. In Sweden there weren't any pellet stoves in 2000; four years later they were going through 40 percent more pellets than all of North America. Davis will be happy to sell you, for about $9,000, a boiler that automatically feeds and burns pellets and will heat your house very reliably and efficiently. It's made in Denmark.

"The Europeans," Davis told me, "look at us as being extremely backward."

Forest Energy Corporation occupies a medium-sized industrial building over which tower huge piles of tawny wood chips. The chips are fed into a grinder that chews them into chunks no more than a quarter inch in length and then, eight tons at a time, into a huge drier that looks like a spinning, horizontal grain silo. When the chunks are down to 8.5 percent moisture it is time for them to be fed indoors and into one of three beefy pellet mills. Forest Energy's electric bill is substantial: about $30,000 a month. The mills heat the wood to just over the boiling temperature of water and compress it into tiny cylinders, about a quarter inch in diameter, that look like something you'd feed a pet rabbit. The cylinders have a glossy sheen because of the heated lignins—natural fibers—that hold them together. After cooling, the pellets are fed into forty-pound bags that are shipped to stores in Arizona, New Mexico, Colorado, southern California. You can buy them for about $3 a bag, which is not a bad deal at a time when prices for oil and natural gas are rising. Forest Energy also grinds up pellets and sells them as absorbent animal bedding as far afield as Texas and Indiana.

The economics of the pellet business are such that stock piles up in spring and summer and flies off the shelves in fall and early winter. Davis was hoping to produce and sell fifty-five thousand tons of pellets in 2005. When I visited it was springtime, and many of them already sat, shrink-wrapped on pallets in white plastic bags, in the large yard outside the plant. During the 2002 Rodeo-Chediski fire the plant was evacuated, and glowing embers were making landfall all over that yard.

Everyone agreed that Rodeo-Chediski was a disaster, but recriminations about it began even before its flames died down and even before President George W. Bush—with a logging-heavy forest-

management initiative of his own that he wanted to promote—
diverted Air Force One to visit some of the evacuated victims. Politicians claimed that environmentalists had in effect caused the fire by
making it too difficult to implement timber sales. Environmental
groups shot back that years of mismanagement by timber and grazing interests had done much more to make the forest flammable than
a few environmental regulations. The rhetoric was heated, the arguments unproductive. A parody version of the Billy Joel song "We
Didn't Start the Fire" that took the environmentalist viewpoint went
the rounds, samizdat-style, on the Internet.

Amid the heat and smoke it was easy to forget that a wildfire,
even in today's altered forests, is not always a bad thing. Within the
perimeter of the Rodeo-Chediski fire there are hundreds of thousands of acres that burned extremely hot and that may take hundreds
of years to grow back into ponderosa pine forests again, if they ever
do. A lack of pine seeds, and a changing climate, may see to it that
they persist as grasslands or shrublands instead. But there are also
extensive areas—many of them areas where logging had thinned the
forest—where the fire burned moderately, leaving behind many living trees. These areas are a reminder that fire can itself do a pretty
good job of restoration in some places and in some conditions. Some
of the healthiest ponderosa pine forests are in places such as the Gila
Wilderness, north of Silver City, where fires, whether ignited by lightning or by people, have burned often. In some places, especially far
from towns, restoration can happen without any tree cutting at all.

The Rodeo-Chediski fire, though, certainly underscored the need
to work on the forests around Show Low and other Mogollon Rim
towns. It was one of the forces that led Forest Energy to expand.
When the plant began operating in the early 1990s it used raw
materials from furniture manufacturers, molding plants, and
sawmills. This feedstock came to the plant fairly dry. Wood that
came directly from living trees, Davis knew, would have a moisture
content of about 55 percent and would require a lot more drying.
But supplies from his other sources were beginning to dwindle as
Phoenix furniture manufacturers lost out to foreign competition.
Davis gambled again.

"I figured that was where the more reliable long-term supply would be," he said. In 2004 the Apache-Sitgreaves National Forest, which surrounds Show Low, was looking for someone to thin thousands of acres of forest a year around communities to reduce the fire danger. The Forest Service was planning to use a new sort of purchasing option, called a stewardship contract, to pay someone to do the work over ten years. Davis put in a bid with Dwayne Walker, who owns a small local logging firm; the only rival bid came from Louisiana-Pacific, the giant logging company. Davis and Walker won. Walker's firm would thin some 150,000 acres of forest over a ten-year period. The idea was to thin primarily small-diameter ponderosa pines from ten thousand to fifteen thousand acres a year around Show Low and other towns. After thinning, national forest staff would conduct prescribed burns to further reduce the fuel load and promote the growth of herbaceous plants. The cost to the federal government would total somewhere between $19 and $91 million over the ten-year period, depending upon the generosity of Congress. By April 2005 fifteen truckloads a day of chips from the thinning project were being delivered to the piles outside the Forest Energy building. They were mainly from trees too small to be used for any other purpose, many of them under five inches in diameter. The trees to be cut as part of the project range considerably larger than that. Some are over sixteen inches in diameter. Those over twelve inches in size go to a sawmill. Many just a bit smaller than that can be sold as posts. Davis takes the leftovers—"the most expensive to harvest, and the stuff no one else wants." The great advantage of turning these trees into pellets is that everything can be used—wood, bark, needles. "You can pelletize just about anything," Davis said. "Cardboard, grass, lint from Levi's pants."

In the first year of work Walker was aiming to thin eleven thousand acres. Davis was employing twenty-eight people in converting the excess into pellets, but he was not sure how much money the federal government would be able to pump into the project over the years. "The only way we can pay for thinning in the long term is to have economically viable businesses that function without subsidy," he said. "We're not going to be able to rely on the government to put money in. We're going to have to do it ourselves."

To have loggers cutting trees on thousands of acres of public land per year, and to have the resulting wood sold for profit, might, but for two factors, sound like the bad old days of overexploitation of western forests. First, the biggest trees are left standing. Second, you can get a hard-core environmentalist—Todd Schulke, to be specific—to praise the project. "I think it's exactly the right way to approach these issues," he told me. "It's probably the right scale. Had that stewardship contract gone to Louisiana-Pacific we would have been very concerned. When people are trying to maximize the amount of wood they can get, then I start to really question it. For Rob, this contract is a ten-year growth contract. It's a huge risk for him. I'd rather have someone like him who can grow into something like this rather than someone who needs all the wood right away. The community-based approach gives us room to work with people."

Schulke expressed considerable hope that work under the stewardship contract can treat most of the fire-threatened acreage immediately around Show Low and other nearby towns. As the worst fire danger around towns is ameliorated, planners will be able to be more liberal in their use of prescribed fire in the wildlands, or even in letting naturally ignited fires burn. He was planning to disagree with Rob Davis about that, but not for another ten years or so. Schulke advocates treating as many acres with fire as possible. In the future, he suggested to me, once the fire danger to human communities has been reduced, it should be possible to have fire sculpt most of the national forest landscape as it once had.

"I suspect that there will be some long-term maintenance needs, especially near communities, but if we're going to maintain these stands in the future it's going to have to be with fire," he said. "If we aren't going to get fire back in, why are we doing this now? Are we going to grow more trees for fiber production? I think it's crazy to think we're going to maintain these forests with mechanical thinning."

He's right, of course; it would take a huge effort to continue to suppress wildfires and to thin huge tracts of southwestern pine forests on a regular basis over the long term. From today's perspective, though, what Schulke advocates seems equally challenging. "We're not going to burn a million acres a year in Arizona," Davis told me, pointing to concerns about smoke, the safety of prescribed burns, and a lack of funding and staff in the state's national forests. Small

pine trees, he said, constitute "a resource I don't like to see wasted if
we don't have to. It's not a waste product. If oil goes up another fifty
bucks a barrel this is going to be one of our best resources. We ought
to get some yield out of these trees."

He knew there was disagreement here. "Some of the environmentalists want to know how to burn a maximum number of acres," he
said. "I'd like to know what the minimum is that we can burn. We
have pretty divergent numbers about this. But at least there is a discussion. We're making a lot of progress."

To thin or to burn? To let as much fire as possible burn in the forest according to its own itinerary—crawling along the ground here,
blowing up into the crowns there—or to bag it instead and feed it
slowly and consistently into stoves to warm peoples' houses? The
question will continue to hang over the Mogollon Rim like the thunderheads that form there almost every July afternoon. For now it is
important that people are going out into the woods to work, and
that they're making decent money, and that the way in which they're
working is more attuned to the ecological realities of the place than
most forest work in the region has been for a long, long time. Somewhere the ghosts of John Muir and Gifford Pinchot are, if not smil-
ing, at least acknowledging each other.

ॐ

FOR TODD SCHULKE, the difficulties of marketing small trees are
more than an abstraction. He is an environmentalist who has
stopped timber sales, but he has also marked his share of trees for
cutting and has served as a board member of a nonprofit group, Gila
WoodNet, that tries to sell their wood.

Schulke grew up in an Iowa farming town but now lives just outside Silver City, which has long been a working town (the documen-
tary film Salt of the Earth, about a miners' strike, was filmed there in
the early 1950s). Silver City is surrounded by huge open pits and has
lived through cycles of the boom-and-bust economies of extracting
silver and copper. In its faded downtown you can see beautiful old
art deco façades on shuttered stores, but you can also see thriving
coffee shops and art galleries that demonstrate how the place is trying to reinvent itself as an artists' colony and retirement haven. At six

thousand feet the climate is mild, the nearby Gila National Forest and Chihuahuan Desert beguiling.

In the 1990s Silver City was hurting economically. Unemployment was in the double digits, and more than 30 percent of the county population lived below the poverty line. Schulke, meanwhile, was trying to figure out what to do in the woods. He'd become keenly aware of the dangers of fire. As it happened, the Forest Service was planning a timber cut in the Gila north of Schulke's home, at what was called the Mill site. Schulke was pretty sure he could shut that sale down—too many big trees were slated to be cut—but perhaps the Forest Service would be interested in conducting a restoration cut instead. It was. Forest Service officials agreed to collaborate with Schulke, and with Wally Covington, to come up with a thinning prescription for an eighteen-acre demonstration site. The project was eventually marked largely according to Covington's ideas.

The marking of the trees to be cut was, Schulke said, a challenge. "We agreed on the general objectives," he said, "but when it came to actually marking the trees we realized a general approach doesn't provide all that much guidance." In the end neither he nor the Forest Service was satisfied with the severity of the mark, and they went back and altered it to leave more large trees.

That was one issue. The bigger issue was that there was practically no one in Silver City who could extract the trees from the forest or do anything with them once they'd been removed. Finally, a man named Gordon West came looking for Schulke. A logger and woodworker, West had recently moved to Silver City from Idaho. He owned a small business named Santa Clara Woodworks and was looking for logs so that he could build cabins.

"Frankly," Schulke said, "I was a little skeptical to have a logger running me down and looking for logs." But their meeting led to a productive collaboration. West, who has a boyish mien and an infectious grin, had the expertise in logging, and in processing wood, that Schulke's little project desperately needed. He also understood Schulke's point of view. He himself had been closely involved with a conservation group in Idaho. "Some loggers say they're environmentalists," he told me. "I actually was, to the point of appealing timber sales. I was not only not afraid of environmentalists; I knew exactly what the issues were and was as interested in addressing them as the Center was."

The two men, with the help of a few other collaborators, set up a nonprofit group with the goal of thinning the forest according to restoration prescriptions and working to develop products from the wood. By early 2005, when I went to visit West, the business was thriving, in a small way. Gila WoodNet's loggers were out in the woods cutting small trees with low-impact logging equipment that was easy on the forest soils. Several local woodworkers were buying small logs from the organization and using them to craft such products as furniture and vigas and latillas—the round ceiling beams popular in southwestern architecture. And Gordon West was busy working out more products.

In his airy shop he showed me his latest piece of equipment, a long metal machine resembling a woodworker's lathe overgrown to a length of about twenty feet. A local machinist had just custom-built it according to West's specifications. The working space in it was occupied by a four-foot-long, five-inch diameter pine log. Crooked, and welted with knots, it looked more easily suited to stoking a campfire than to supporting a house roof.

To West, though, logs of that sort represent the future. He'd had the new machine outfitted with an integrated bandsaw, drill, and laser sight. It allowed him to take small logs and saw them, drill holes, and install fasteners in them so that they could be used as interchangeably as two-by-fours. "My plan is to make interchangeable components with natural logs," he said. "With this equipment, you can turn these logs into trusses that you can use as precisely as milled logs, even though they're all random and bent. Do you know that this little log is probably as strong as a five by five post? If you can use a natural log like a sawn log, then you save a lot of energy and have a lot less waste. And in a lot of cases it's more aesthetically pleasing." And messing with the new equipment, he admitted, was not a bad way to spend the day. "I'm going to have a lot of fun with this," he grinned. "Ah, new toys."

This machine was only one of his new toys. He was also working on a new product called Chipcrete, which will turn wood chips into durable bricks that can be used like cinder blocks. He was building, on a wide-tired Mercedes truck chassis, a new yarder designed to haul small logs from the woods with minimal effect on soils and plants. In fact, West was spending so much time on this sort of product-development work, and on such administrative tasks as writing grant proposals and overseeing Gila WoodNet, that he did not have much time for his true heart's desire, the work that started it all. There was a cabin frame in progress in the yard outside. It was built of peeled ponderosa logs, none of them larger than ten inches in diameter. West had clients who wanted him to build such cabins for them, but he seldom found time to work on them.

Progress in the woods was also slow. Gila WoodNet's original goal was to thin about five hundred acres a year. The Forest Service had done an assessment that showed that ten thousand acres of ponderosa pine forest in the Silver City area stood to be thinned to reduce fire danger. Five hundred acres a year would spread the work out over twenty years—a practical planning period for those trying to figure out how to structure careers or manufacture and market products. There were not yet enough markets, though. Rob Davis, who grew up in the area, had suggested that Gila WoodNet could sell wood chips for heating Fort Bayard, a nearby state hospital, and the state had, in fact, allocated funding for the conversion of one of the facility's natural-gas boilers to wood. But there were political difficulties, and the state wasn't sure it wanted to continue running the place. By the fall of 2005 it remained unclear whether the conversion would ever happen. It was frustrating. Thinning was proceeding at a snail's pace; in the meantime, the state kept buying fossil fuels to heat the hospital.

"Two years ago Fort Bayard was paying $140,000 a year for natural gas," West told me. "It's probably more now. We figured that it would pay $100,000 to $120,000 a year to burn wood biomass instead. That's a savings, but what's most important to the community is that the state will spend that money in Grant County rather than paying it out to some company elsewhere, probably in Texas. Almost all of the money will stay here. That's probably worth more to the state than $20,000 in savings."

Gila WoodNet needed to find some market like Fort Bayard or Chipcrete—or even truly low-value products such as mulch—for the smallest of the trees harvested. It was easy, West said, to sell the larger trees that could make vigas or house logs, but only selling those did not compute. "That 20 percent of the material that goes into vigas, latillas, and so on generates about 80 percent of the income, and the 80 percent that goes into chips or biomass generates about 20 percent," he said. "But you don't want just one. The ecological considerations of the forests require that you deal with the mass of material that can go into biomass, but the economic health of the community requires the higher-value products. You can't neglect one or the other. We can't have the 20 percent I need for cabins if we don't also handle the 80 percent."

From the perspective of the modern global economy, Gila Wood-Net had, in a sense, deliberately hobbled itself. West and Schulke left more of the valuable larger trees standing than Covington's thinning prescription would have, and they wanted the project's production to mirror only local demand. They didn't want to be selling vigas and cabins to customers in Santa Fe; they wanted to sell them locally. Santa Fe had its own forests to thin. "Big industry isn't where it's at," West said. "I'd like to see dispersed small industries. That's where it's at. You shouldn't ignore the small solutions. They're perfectly good and will last. The more dispersed an industry is, the more stable it is and the less likely to crash."

That all had a nice ring to it, but it also sounded really idealistic, and it was difficult not to wonder how realistic a business model it represented. West might be looking to sell products only in the Silver City area, but the products he competes with there are those that have come out in front in a highly competitive global marketplace. Whether or not Gila WoodNet will succeed in the long run will depend on the willingness of local customers to buy its products, even if they may end up costing a bit more than products shipped from elsewhere. The success of businesses such as West's and Davis' may well hinge on the global economy. If the price of oil, and hence the cost of transportation, continues to rise, then such local- or regional-scale enterprises will start looking like pretty good bets.

"I think Gordon's going to make it," Dennis Becker told me. Becker is a forest economist at the University of Minnesota who as a Forest Service researcher studied small-scale utilization projects in the

Southwest in recent years. "Gordon's a smart man. If you can capture and supply local demand, that's great. There are some products in the home building market that he can provide, whether roundwood logs, moldings, maybe non-load-bearing two-by-fours for interior construction. He knows what's needed in the local economy, how to work within the local building codes, and he thinks he can outcompete other products with his local products because he'll have lower transportation costs. Maybe he can even sell them for a slight premium as local products. It's a good model. But even so, the cards are stacked against him. The global economy is overwhelming."

The equation seems simple: do good work in the woods, sell high-quality products, make a good living, persist. Will it balance out? Check back in a few years. "I see this as creating a new culture in the woods," West told me. "What we're trying to accomplish comes first and making money comes second—though it does come. But you don't care so much about making a lot of money if you're already doing what you want to do."

⁓

GIVEN ALL THE HULLABALOO about forest restoration—ecologists and forest managers hobnobbing with governors and senators, the hot glare of media attention, fingers of blame wagging this way and that whenever a big blaze burns—it is easy to forget that it's taken a long time for the forests of the Southwest to reach their current state and that it will, by the same token, take a long time to repair them. The U.S. political system, like its electorate, likes quick solutions. It goes for the quick conflagration, not the slow smolder. It does not like to be told that solutions to anything will take decades or perhaps generations. Yet that is exactly the case with forest restoration. Even if all the thinning required to return the density of southwestern ponderosa pine forests to reasonable levels were to take place next week, it would still take many decades or even centuries before all those skinny pines could grow to the stature of the big old yellow pines that were cut long ago. In fact, thinning will take many years, and more seedlings will continue to grow as it takes place. In the interim some dense stands of pine are no doubt going to burn as intensely as those toasted by the Rodeo-Chediski fire. Some areas will be thinned according to Wally Covington's ideas and some to Todd Schulke's. On some thinned tracts the Forest Service will never get around to conducting prescribed burns, and they will again grow dense with small trees. Some areas will be thinned and burned and will look spectacular. Some will never be treated, except perhaps by a lightning strike, and will look equally spectacular. Some will be thinned with care and will end up looking like tree farms.

The result is going to be a patchwork of areas that have burned ferociously, of areas that have burned at low levels of intensity, of areas that have not burned at all. And this crazy quilt of ecological circumstance will likely mirror an economic patchwork. Americans have a tendency, as Dennis Becker put it, to look for a "silver bullet." For a problem as ecologically, economically, and politically complex as forest restoration, however, there is no such thing. The solutions to the problem are manifold, and they are all incomplete. They will have to work at different scales at the same time. Yet there's no reason why they can't. Consider the medical metaphor again. Americans are happy to visit internists who have a small family practice; we also want to have available highly centralized hospital complexes equipped with the latest CAT scanners and oncology centers. We want intimate, boutique-style care; we also want the sort of technology-centered efficiency that is achievable only when patients come in quantity. If the health-care economy can function—not always perfectly, to be sure, but effectively—at these different scales, perhaps the forest-care economy can too. There is no reason why Rob Davis, processing the wood from ten thousand acres a year, cannot thrive at the same time as Gordon West, working on a few hundred.

"We don't need to have every approach at every level," Becker said. "You have to have the Gordon Wests, but you also have to have something that works at a regional level." Despite a long history of poor decisions made with too little consideration for ecological realities, there is no reason that there cannot be a diverse and responsible forest-products industry in the Southwest. Indeed, Becker suggested that the sheer lack of logging and processing infrastructure in the region might be an advantage. The industry is in the process of reinventing itself practically from scratch, without a lot of established players and vested interests. "In the Southwest we get to have a say in what the industry will look like," he said. "I think we're actually in a really unique situation because of that."

What might initially seem a liability, then, may turn out to be an opportunity. When you view ecological restoration through an economic lens, as any entrepreneur must, what it boils down to is that its ideal result is an expansion of possibility. Rob Davis and Gordon West today are stuck with lots of small trees and few big ones. In a global marketplace, they're left with a pretty poor hand, which is not to say they can't play it well. But if restoration succeeds on a large scale—if today's ecologically minded foresters can succeed in growing a lot fewer ponderosa pine trees, but bigger ones—future generations will have an expanded set of opportunities. They may have to choose between turning old-growth trees into lumber or pellets—or letting them stand. Todd Schulke, for one, views his work as culminating in such decisions. He'd like the forest's possibilities to approach those it had in the past. "If my descendants decide to cut old-growth trees again," he said, "there'll be lots of them."

The real end result of forest restoration, then, is as much about people as it is about trees and bunchgrasses and fire. It is about ensuring that future generations have a generous rather than a constrained set of possibilities. They'll have the ability to do more with fire, and with the chainsaw. Maybe a great deal of time and energy and money will be spent on forest restoration in coming years simply so that southwesterners in eighty years, or a hundred, can cut big trees as rapaciously as the early loggers did. Maybe the new restoration economy won't lead to anything sustainable at all, at least not in the long run. As prices for fossil fuels rise, maybe a hunger for wood pellets will lead to all sorts of cut-and-run logging.

The best we can do, though, is work toward what looks sustainable to us from our constrained, time-bound viewpoint. What that amounts to is bringing together the ghosts of John Muir and Gifford Pinchot. To do so would be to recognize that the viewpoints each held are, by themselves, as incomplete as half a magnet. Throughout his career Pinchot advocated eliminating fire from the forests, never realizing that natural fires serve purposes no human management could. He never recognized that a knee-jerk policy of trying to get rid of every small fire led inevitably to larger and more destructive fires, to greater and greater deviations from a healthy ecological norm.

And Muir? Muir wrote rapturously about the complexity of wild places but never addressed in his writing where the timbers that built his own house came from or even where his meals originated. He was the perfect precursor to the contemporary leave-no-trace back-packer who heats foil-wrapped dehydrated meals on a butane-fueled backpacking stove. Muir suggested a beautiful, respectful way to visit the wilderness, but never addressed the topic of living anywhere, not in the sense of actually *making* a living.

Pinchot's philosophy could work only if nature were as simple as a laboratory experiment; Muir's works only on a local scale because it relies on rigorously excluding places that aren't wilderness from consideration as healthy, wild places. No wonder the two of them had a good time together on the South Rim—they needed one another as the two legs of a stepladder do. In recent decades the practices of land management in the Southwest have swung wildly and litigiously back and forth between Pinchot's ideas of human use and control of the environment and Muir's hands-off ideas. Amid all the conflict it's been easy to forget that a magnet is a useful tool in navigation and that every stepladder goes somewhere. Maybe it's time for us to climb on up.

Where might we arrive? Somewhere a bit messy, is my guess. Someplace that doesn't quite mirror any one person's image of perfect forest management, but in the aggregate represents a larger set of ideals about how people can live in nature without wrecking it. The middle way is the slippery high ground between well-worn ruts, and the practice of restoration, as Bill Jordan suggests, is a more promising place than most to look for it. Where it takes us might just be a place that looks like home. Think about what the word *home* means. Home is four walls, to many of us. Shelter from the world outside. A warm hearth. Lights that work, water that runs, heat, a full refrigerator and cupboard, the systems that make you comfortable. The place where you feel safe. Where the heart is, they say.

That's true, but it's also sentimental claptrap. Home is also where all the other, messier, organs reside: the bowels, the genitals, the mouth. Home has a lot of thoroughly unsentimental aspects. Home is the place where, if some of those systems don't work, you have to fix them. In a country as footloose as the United States it might make the most sense to define home as the place where there is no running from responsibility. On a fine sunny afternoon you might congratulate yourself on how organized and comfortable your life is, and then that night the baby won't stop crying or the cat throws up on the carpet or the bathtub drain clogs and you remember that at 2 A.M. on a cold morning there is no escaping the reality of home into the dreamlike comfort of some ideology or abstraction. Home, you realize, is a long-term relationship. Home is life with the bark on, the place where you have to make the hard decisions. Home is where you choose, at some point, to have the plumber come and rip out the wall to replace the leaky old valves. Home is where you make the call that, yes, it is time to put an elderly relative in a nursing home. Home is where you make sacrifices, out of caring, for the future—for your children or for their children.

John Hogan, one of the coordinators of a volunteer task force that has done a great deal of erosion-control and revegetation work in Los Alamos, New Mexico, since a runaway prescribed fire destroyed more than two hundred houses there in 2000, once told me: "People say, 'I used to see dense forests, and now all I see is fuel.' It's the kiss of death, a loss of innocence."

Loss of innocence is right, but then what kind of life would it be otherwise? An adolescent one, that's what kind. Growing into adulthood is all about beginning to comprehend the kiss of death, and innocence is something that exists to be lost. Despite what John Muir wrote, innocence does not go far in managing the complexities of adult life.

Add, then, to the list: home is where you do not back away from the blood and grit of existence. Home is where you make some compromises, because anything that lives must. Home is where you embrace the fire and the sawdust, and you come away ashen and complicit and very much alive. There are stumps and there is smoke, and sometimes the fire is corralled by human intentions and sometimes it gets away and does what it wants because it, too, is alive and at home.

‰

IN FLAGSTAFF TWO POTTERS IN THE 1980S orchestrated the construction of a pair of enormous kilns. One, Don Bendel, was a professor of art at Northern Arizona University. The other, Yukio
Yamamoto, was a master ceramicist from Japan. Together the men
oversaw the construction of two tozan kilns on the university's cam-
pus. "Tozan" means "east mountain," and it refers to the region
where pottery kilns were built in classical Japan. They were traditionally of two varieties: noborigama, the climbing kiln, and
anagama, the cellar kiln. Both were long and built to rise up hillsides
and were stoked with wood—lots of wood.

In Flagstaff, Yamamoto and Bendel built one noborigama kiln,
with multiple chambers, and one smaller anagama, with a single large
one. They were the first tozan kilns in the western hemisphere. For
twenty years now ceramics students have fired their work in them.

One evening my wife and I went to see the noborigama kiln get
fired up. The twin kilns sit under an open metal-roofed shed, in the
shade of pine trees. Constructed of bricks and of patches of variously
colored cement and shaded with soot, the kilns look ancient. Firing
them takes a week and a whole village of ceramics aficionados. First,
the stacked rooms of the kiln are closely filled with fresh pots, vases,
and sculptures: the moldings of myriad hands and fingers working
countless hours. Then the burn begins. For weeks beforehand the
students have been chopping great heaps of pine logs, thinned from
the forests around town, into strips of kindling. When the piles of
kindling have grown to vast proportions it is almost hard to see the
kiln among them. Then it is time to ignite the fire.

They light it in the evening, feeding the ready kindling into the
kiln's arch-shaped maw. From there the kiln's chambers stair-step up
forty or fifty feet. The sky grows dark, but the arched entrance comes
to glow orange and then red. The sticks are fed in, one by one, into
the night. Slowly, slowly, the great mass of bricks heats up. Patience
is as great a virtue for potters as it is for foresters.

When the night sky is fully dark we go home, but the students
and other helpers will keep up the work, around the clock, for
another week, feeding the kiln, tending the fire in the heat of the
afternoon and the cold of the night, reducing the great piles of pale
wood to ash and smoke.

And to something more. After the fire is out it takes days for the kiln to cool down, but once it does the pottery comes out. The great heat of the noborigama, and the abundant ash from its wood fire, renders beautiful dark streaks on the glazed surfaces. The potters never quite know what vessels will look like when they've been fired. Some are spectacular. Some crack or shatter.

A friend gave us a bowl from the noborigama as a wedding gift. The clay is the deep-hued rose of ponderosa bark. Covering it all, save the base, is a green glaze that runs from olive to palest chartreuse, the color of a pine canopy washed out by bright midday sun. And all over the bowl are dark black streaks, thin as pine needles, etched by fire and ash, impossible to anticipate, unrepeatable.

The bowl is earth transmogrified by fire, a symbiosis, shaped in equal parts by the forces of nature and by the creativity of human hands. It holds the memory of fire and of an equal covenant between people and nature.

A pessimist might point out that the wood that fired it, removed from the forest, hardly makes much difference to the wildfire danger around Flagstaff—it comes from a few acres at best.

But then an optimist, or anyone who sees promise where others see only problems, might reply that the making of Communion wafers makes scarcely a dent in the global supply of wheat.

4

The Voyage of the Moon-Eyed Horse

*I*t was the grass that really got me. All the rest I could grasp: the billowing red rock desert that seemed to stretch to infinity all around, the blue slate of the sky only slightly marred by white contrails, and yes, the great concrete plug of a dam backing up the flat immensity of water that we could see to the east. To try to comprehend the height and weight of all that water—the dam was more than seven hundred feet high—gave me a feeling of vertigo, but still I could understand it. Glen Canyon Dam was physics interrupting gravity. It was cold and hard and precise. It was, if you had a long enough line, quite literally fathomable.

But the grass? That didn't compute. Why would anyone plant and so visibly, so carefully mow a semicircular field of grass in the middle of a desert where it only rains six inches per year, down between the base of the dam and the turbine house, where no visitor could ever go?

The grass was there, Harry Gilleland told us, precisely because this was a desert. "The public likes to see something growing in the desert," he said. It was 1990. Gilleland, retired from full-time work with the Bureau of Reclamation and our tour leader, had a down-home accent that I found out later stemmed from Arkansas. "That grass is a beautiful sight that shows the beauty of the desert when it has water."

I was touring the Southwest with a group of graduate school students in environmental education, and we were full of piss and vinegar. We wore sandals and ate hummus and beans and slept under the

stars. We craved the wild and the untrammeled. We'd been saddened
to hear how Lake Powell had drowned almost two hundred miles of
the Colorado River's fantastically scenic Glen Canyon. We'd been
reading our Ed Abbey, who in his books *Desert Solitaire* and *The
Monkey Wrench Gang* fantasized about blowing the dam to bits and
reclaiming the canyon. We'd heard how the radical environmental
group Earth First!, taking a cue from Abbey, had in a famous publicity stunt unrolled a long mock crack of plastic down the dam's face
back in 1981. We had a cassette tape of songs by a seventy-year-old
Arizona folksinger named Katie Lee who'd made a crusade of casti-
gating the dam and the bureaucrats and politicians who'd built it.
The songs had titles like "They Crucified My River" and "The
Wreck-the-Nation Bureau Song." We hated the dam, quite simply,
and all it stood for: the blind faith in historical progress and in the
prowess of engineering, the notion that nature stood to be improved
by humans, the rigorous allegiance to what we considered the highly
dubious goal of sending electricity to Los Angeles and Phoenix and
Las Vegas to power ever more air conditioners. We were on our way
to a backpacking trip to Rainbow Bridge, a natural wonder that once
had been idyllically difficult of access but now had the artificial lake
lapping virtually at its bases. Thanks to Lake Powell, you could practically drive a houseboat to it. We could not help but view that as a
diminution. We had, in short, met the enemy, and it was Glen Canyon
Dam and the agency that had built it. And Harry Gilleland was the
designated emissary from that foreign land to ours.

   He was unequivocally of the old school, but perhaps he was playing things up a bit simply to ride us. "The construction of the dam is
a monumental achievement," he intoned, as if parroting the film
loop playing in the visitor center. "Why, it cost $1 million just to out-
fit the elevators with solid-state circuitry." Did the constant buzzing
of the giant transformers that sent electricity humming on down to
the lowlands ever bother him? we wanted to know. No, he said, "it's
music to our ears because from that we get our pay." Wasn't the dam
built in soft and unsuitable, even dangerous, sandstone? someone
asked. We'd been doing our research. Hadn't the dam almost failed
during the great flood of 1983? No, he replied. Navajo sandstone
"wasn't a weak rock," he said, it was "an unusual-type rock." What

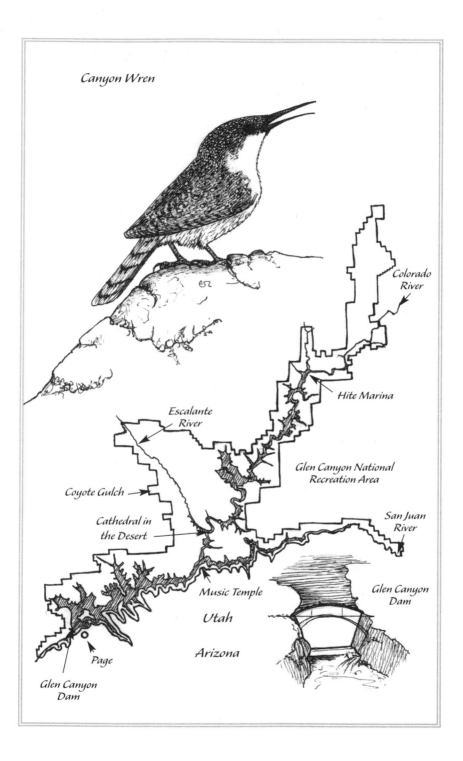

Canyon Wren

Colorado
River

Hite Marina

Escalante
River

Glen Canyon National
Recreation Area

Coyote Gulch

Cathedral in
the Desert

San Juan
River

Music Temple

Utah

Glen Canyon
Dam

Arizona

Page

Glen Canyon
Dam

about other development in the area? What should happen to the surrounding country, Harry? There was an "endless supply of coal, an endless deposit" to the north up on Utah's Kaiparowits Plateau, and to the south on Black Mesa, down on the reservation, was another "endless" supply. Why, this heart of the Colorado Plateau was practically an electricity factory, exporting its product on copper cables all over the southwestern quarter of the country and in return raking in the receipts. It was a humming, efficient machine, and the dam and the nearby town of Page, Arizona, were integral parts of it.

The gates of the dam, we learned, had been closed in early 1963, at the high-water mark of the American century, before President John F. Kennedy's assassination, before Vietnam heated up, before Watergate, the energy crisis, global warming—before a whole lot of inconvenient questions came to be asked. Its smooth concrete face swooped upward so cleanly, a pale-faced plug in a red canyon—and it had all been built in only a few years! It was progress manifest. As was true of jetliners and skyscrapers and '57 Chevies, there was something sexy about it.

We toured the visitor center and the power plant hidden deep between canyon walls and even crawled inside an empty turbine usually filled with a chaos of pressurized water, which gave me another sense of vertigo. I should confess here that I was feeling my oats. Against the regulations of both my school and the Bureau of Reclamation I slapped an Earth First! sticker inside the eerie, curved tube. I don't know whether some Bureau employee found it or if it remained undiscovered and ended up dissolving, molecule by molecule, and finding its way down into the Grand Canyon after having done its small part to supply the West with cheap and abundant electricity.

We ended our visit on top of the dam, looking out over the cobalt blue expanse of Lake Powell or, as the promotional literature inside had proclaimed, "America's Natural Playground." The playground was reined off from the dam by a floating boom. "We have to take the threats to the dam seriously," Gilleland said, referring, we figured, to Edward Abbey and his buddies. There were, he acknowledged, people who thought the dam had ruined something. They just couldn't see the grandeur he did.

I'd been reading too. Isn't the reservoir filling up with sediment?

I asked. How long will it take before it fills up? I was wearing a beat-up cowboy hat, and Gilleland flicked its brim playfully with a finger. "You're not supposed to ask that," he said. Eventually, though, he did come around to an answer. One hundred and forty-five years, he said. Enough time that neither bureau veterans nor young punks like us would need to worry about it.

⁊ͻ

IT WAS ON ANOTHER LOVELY SPRING DAY, almost precisely fifteen years later, that I found myself motoring to a place called Cathedral in the Desert on the inaugural voyage of an ungainly vessel named the *Moon-Eyed Horse*. The boat, about twenty feet long, consisted of a rectangular platform mounted atop two pontoons, each welded together out of steel tubes just longer than oil drums. Its top speed was stately: three miles per hour. Its vinyl seats and weatherproof carpeting and flimsy-looking chrome railing gave it the appearance of a cheaply built floating rec room. You could spill beer or fish guts anywhere and not worry about the consequences. The desert was the country we'd seen from the dam long ago, the channeled and domed slickrock badlands that extend from Page far northeast into Utah. And the Cathedral? That was merely one of the great icons of Glen Canyon, a place that all the old river runners spoke of in hushed tones, a place that fifteen years earlier we and almost everyone else had considered as irretrievably lost as Atlantis.

The wind blew sharply from the north and ruffled the surface of Lake Powell, which had a stark, even bizarre sapphire hue against the red and cream of the sandstone walls. The walls formed a labyrinth of corridors that plunged straight down into the depths, ungentled by beaches or even steep talus slopes. There was soaring, smooth rock, and there was flat water: a landscape all warp and weft, as reduced to basic verticals and horizontals as a stick of butter on a serving dish. The gently undulating walls reached hundreds of feet above and pinched together ever more tightly, shutting out the sunlight.

The captain of our modest vessel was Andy Peterson, a muscular young rock climber from Salt Lake City who was all blond hair and blue eyes and big grin. He'd just begun working for a little outfit called the Glen Canyon Institute (GCI). The passengers were so bundled up in parkas and gloves and woolen hats and sunglasses that we looked more like dogsled than boat passengers, but along with me they included five retired schoolteachers, all women, from the West Coast. They were avid hikers and campers, and they were all true believers. What they believed was that building the dam had been a terrible, rotten idea.

We knew we were getting close when we began passing the trunks and branches of skeletal cottonwood trees. Drowned when the reservoir filled, they reached out of greenish water from ghostly depths and stood stark against the sheer walls, well-preserved hulks exposed now by the sharply dropping water levels. The Southwest had been in serious drought for the last five years, and the lake's surface stood 140 feet lower than it had back in the early 1990s. The drought was the reason we were able to visit the Cathedral, and it was what fueled the GCI's desperate hope that it, and many other places in Glen Canyon, might be spared from being flooded once again.

We could clearly see just how much water had been lost, for the rock reaching up from the surface to a height of 140 feet above us was bleached white, thanks to minerals in the lake water, with what everyone referred to as the "bathtub ring." The surface stood just a bit lower than it had in 1969 when, during the lake's long period of filling, the writer John McPhee brought the conservationist David Brower and the Commissioner of Reclamation, Floyd Dominy, here together on a boat trip.

"Putting water in the Cathedral in the Desert," Brower said then, "was like urinating in the crypt of St. Peter's."

"We destroyed something really beautiful," Dominy admitted. "But we brought in something else. . . . What we got is beautiful, and it's accessible." Lake Powell, he said, was "the most beautiful lake in the world."

"I just wish you could hold the water level where it is now, Floyd," Brower had said then. Almost four decades later, that was the GCI's immediate goal too.

We made another turn, and the rock walls pinched together yet more tightly. It was impossibly tight, it seemed, but no—the passageway continued. The boat spun to starboard, and there before us was a deep chamber, almost a cave, a long-stretched alcove carved like

the temples of Petra from rose red rock. A tiny waterfall thirty feet
high spun its way down the far wall from the bottom of a deep, sinuous notch eroded into the alcove. We ground to a halt on a beach of
golden sand, Peterson cut the motor, and silence descended on us.

The place was dim. The blue of the sky was reduced to a narrow,
crescent-shaped slit. Light spilled down from sunlit sandstone walls
hundreds of feet above and filtered, lambent and dusky, into the
chamber before us. We disembarked and tied up the boat.

We could hear the silver tinkle of the cascade, whose water pooled
crystalline in a wide sandy basin and, from that, ran toward us in a
clear stream. Crossing it, we walked onto a golden dune of fine, dry
sand and looked up toward the thin sliver of sky and the high, sunlit
cliffs that enclosed it as if forming a natural chancel window.

It *was* like a cathedral, an airy chamber at once confined and
soaring, a womblike place, almost subterranean, that was absolutely
of its own rocky element but that also pointed inexorably toward the
heavens from which the light, in its soft descent, bounced off the glis-
tening walls. The place had a curvaceous grace to make a Le Corbusier
weep. Countless tiny seeps wept droplets, like dolorous miracles, from
barely visible creases in the smooth rock. It did indeed seem, as a
writer for *Audubon* had written back in 1966, "Nature's altar,
lighted by a window to Heaven."

Then we heard the growling of the motors—one, two, three of
them, grinding their way through the passage we had just navigated.
They grew louder, and within minutes a dozen new visitors, young
men and women, were hopping out of the powerboats and milling
among us, laughing, joking, smoking cigarettes, playing with the
three dogs that, still wearing neon orange life vests, had bounded out
of the smallest of the boats.

Life preservers for dogs? Who knew?

Our quiet reverie was over. We were freezing cold, and it would
take the slice of direct sunlight high on the chamber wall a long time
to work its way down to us. We cast off from the beach and back-
tracked our way past the ghostly trees.

"The past is never dead," William Faulkner wrote. "It's not even
past." His truth is about as self-evident in the depths of Glen Canyon
as it is in the depths of Mississippi. Those trees from another era

remain, reminders of the people who walked under their trembling leaves in summers long past, people like Ed Abbey, and Katie Lee, and Wallace Stegner, and Georgia O'Keeffe, and the photographers Eliot Porter and Tad Nichols and Philip Hyde and Barry Goldwater. Nonetheless, it was manifestly clear as we putt-putted along that that particular past was also irretrievably gone. Where hundreds of people rafted Glen Canyon *each year* before the dam was completed, hundreds *a day* might easily visit the newly exposed Cathedral now in their powerboats. If the Cathedral was going to stay out of the water, I suspected, the Park Service was going to have to moor one of its floating toilets just around the corner. Could the interpretive signs be far behind, and the concession where you could warm up with a cappuccino while renewing the rental of your life jackets, or those of your dogs?

A powerboat was coming up the rock corridor toward us. Its four occupants waved vigorously. Was something wrong? Peterson slowed our boat and spun the wheel. What was it?

"Cathedral?" the man at the wheel yelled, then pointed straight back toward the chamber.

"Yeah, Cathedral," Peterson yelled back, a bit wearily, and gave a thumbs-up.

୬ଡ଼

IT WAS LIKE BUYING AN OLD HOUSE, tearing out its antiquated plumbing system, and starting from scratch with new pipes and pumps. In the American West the twentieth century was the century of reclamation, when the federal bureau of that name replumbed the region's rivers on a vast scale to provide water and electricity and flood control to a swelling population—and, cynics would say, to provide jobs for itself and fat campaign contributions to the politicians who authorized its projects.

At the beginning of the century no river was more in need of replumbing, as the boosters saw it, than the Colorado. It drained an area the size of Texas, including large stretches of the high and snowy mountains of Wyoming and Colorado, but it was wildly erratic. Its flow fluctuated from trickle to torrent, from fewer than a thousand to more than three hundred thousand cubic feet per second. "It is a runaway, an outlaw, a death-dealing demon, striking terror in the hearts of those who patrol the levees," wrote an observer of the river's lower reaches in 1926.

Attempts to divert the river's water for agricultural use in Southern California began around the turn of the twentieth century, but extensive and coordinated development didn't occur until the 1920s. In 1922 the seven states through which the Colorado and its tributaries flow signed the Colorado River Compact, which was brokered in a series of meetings by Herbert Hoover, then secretary of commerce. The delegates effectively split the U.S. portion of the river into two segments at Lee's Ferry, the historic river crossing that sits just below the Arizona-Utah border. The upper basin included Wyoming, Colorado, Utah, and New Mexico, the lower basin Arizona, Nevada, and California. The delegates agreed that the upper basin could use 7.5 million acre-feet of Colorado River water each year and would supply an equal amount to the lower basin. (One acre-foot is the amount of water that will cover a football field to the depth of one foot; it is generally considered to be sufficient to meet a typical American family's needs for a year.) Mexico, where the river debouches into the Gulf of California, was largely ignored, as were the claims of Native American tribes on the river's water. Not until 1944 did the United States formally agree to provide Mexico a consistent 1.5 million acre-feet a year, and most Indian claims are still unsettled today. Not until much later in the twentieth century did it become clear that the compact had been signed during a remarkably wet and remarkably boosterish period when it was easy to develop overly optimistic beliefs about the river's volume. The state delegates who met with Hoover believed that the river could provide some 17.5 million acre-feet a year; hydrologists now believe that the actual long-term average lies between 13.5 and 15.1 million, far less than the established claims.

The agreement wasn't worth much without the infrastructure to implement it, and the Bureau of Reclamation set to work on that with a vengeance. It completed its first truly big waterwork on the Colorado, the monumental Boulder Dam (later renamed Hoover), on the Arizona-Nevada border in 1935. The dam trapped high flows in Lake Mead, allowing for predictable and controllable releases of water into the river's lower reaches. Other dams followed, both upstream and down. The river was tamed. Downstream towns such as Yuma, Arizona, were not in danger of flooding anymore. The lower basin states had predictable supplies of water even in dry years. Boulder Dam, said President Franklin D. Roosevelt at its dedication, was "altering the geography of a region." It was no overstatement. The dam watered lettuce and alfalfa and tomatoes in California's Imperial Valley and lit up the streets of the Southwest's burgeoning cities.

But it wasn't enough. The early benefits of replumbing the river almost all went to the lower basin, which had far more people and a climate much more suitable for irrigated agriculture than the mountainous upper basin. Lake Mead, by far the largest reservoir, could supply water and electricity only to the three lower basin states. The Colorado River Compact might look good on paper, but any politician knew that it *was* only paper. With populations in Arizona and Southern California continuing to boom, it was far from inconceivable that those states might grab more and more of the Colorado's water. Furthermore, building a dam on a river that carries a great deal of sediment begets a stream of logic almost as inexorable as the river itself. The river's sediment will settle in the stilled water, reducing the reservoir's storage capacity and its useful lifespan. With Hoover Dam built, the Colorado's abundant silt began to pile up in Lake Mead. Dam engineers began to dream of building another dam upstream—or more than one—to help it last longer. What the states upstream needed, clearly, were their own reservoirs.

In 1946, with the soldiers and sailors home from Europe and Japan, the Bureau of Reclamation thought it was time to act. It released a report, *The Colorado River: A Natural Menace Becomes a National Resource*, that called for dozens of water-storage projects on the upper river and its tributaries. When it was refined in the ensuing years, the bureau's proposal called for dams at, among other locations, Glen Canyon; Bridge Canyon, which is in the depths of the Grand Canyon; and Echo Park, which happens to lie along the Green River within the boundaries of Dinosaur National Monument in far northwest Colorado.

Conservationists were outraged less by the idea of inundating

Glen Canyon—which was not a part of the National Park system—than by the idea of flooding national parks and monuments. The Sierra Club, a small California hiking club just testing the waters of national politics, initially didn't object to the dams, but when David Brower became its executive director in 1952 he came to vehemently oppose the building of dams within parklands. After all, club founder John Muir had, near the end of his life, staked his political capital on a failed effort to prevent the damming of Hetch Hetchy Valley in Yosemite National Park. If Dinosaur and Grand Canyon were not safe from the bureau in the 1950s, what park would ever be safe in the future? But Glen Canyon? Well, that was no big deal. It was important, the Sierra Club's directors believed, to be reasonable. At a congressional hearing in 1955 Brower said that the Sierra Club "would not have any objection to a Glen Canyon Dam." Brower, who was a skilled alpinist and far more inclined to hike *up* into scenery than float *down* into it, hadn't been to Glen Canyon. When the Colorado River Storage Project Act passed in April 1956, the proposed Echo Park Dam had been excised from it. Bridge Canyon Dam was not included. The legislation did, however, authorize the construction of a dam in Glen Canyon.

Brower found out too late what had been lost. Not until the dam's cement blocks were rising high did he finally float Glen Canyon. He visited the Cathedral. He stopped in at Music Temple, the great natural amphitheater so named by the pioneering Colorado River explorer John Wesley Powell for its astonishingly perfect acoustics. He saw huge, sweeping walls of Navajo sandstone, a thousand feet high, across which ran long vertical draperies of black and umber desert varnish deposited over the course of many centuries by running water. He saw gardens of moss and liverworts and monkey flowers hanging hundreds of feet overhead where water seeped out of the rock. He saw inviting sandbars tacked to the steep canyon walls by groves of lush green trees, and springs limned with trembling cottonwoods. He saw ancient cliff dwellings. He watched great blue herons and beavers, heard the gently descending song of the canyon wren. The scenery, he realized too late, matched—no, exceeded—that of almost all America's national parks. In addition, the experience of the place was perfectly democratic. Unlike the

Grand Canyon downstream, with its adrenaline-pumping rapids that demanded considerable skill of boaters, Glen Canyon was peaceful. The river ground its way through thick layers of soft sandstone that didn't form rapids. Anybody could float down it, as Abbey had, in the flimsiest of vinyl rafts. Boy Scouts and grandmothers could go— or, rather, *could* have gone.

"I didn't know what was in Glen Canyon," Brower said later. "That was one of the bitterest lessons I ever had."

Only six months after the legislation passed, the blasting of the canyon walls began. The Bureau of Reclamation moved into far northern Arizona with a vengeance. The Navajo traded 16.7 square miles of land above the dam site for oil-rich lands along the San Juan River in Utah. The bureau set up a construction camp there, and within a few years twenty-three hundred workers were living in the new town of Page amid the pink sands. Like Hoover Dam before it, Glen Canyon Dam was from its inception all superlatives. It came to rise 710 feet high, almost reaching the canyon's red rims. It took into itself one hundred and thirty thousand tons of steel, twenty thousand tons of aluminum, and five thousand tons of copper, as well as five million barrels of cement and ten million cubic yards of aggregate that was mined from the bed of Wahweap Creek just upstream— enough material to build a four-lane highway from Phoenix to Chicago. At the peak of the construction workers were laying six hundred tons of concrete per hour. The refrigeration plant needed to cool the concrete as it set was large enough to make six thousand tons of ice each day. The eight giant turbines in the dam's heart could produce up to thirteen hundred megawatts of clean, renewable electricity. And all that material, when it was finally put together and completed in 1966, was enough to back up a reservoir that when full would stretch 186 miles and lap at 1,960 miles of shoreline. Lake Powell would store twenty-seven million acre-feet of water. It would consist of two years' worth of the Colorado River stilled and held in place and made to serve the desires of humans, which were, as the Bureau of Reclamation and the politicians saw it, the generation of electricity, the storage of water for the upper basin states, and the provision of motorized recreation on the reservoir.

The first great milestone in the progress toward those goals came on January 21, 1963, when bureau workers sealed a diversion tunnel that had allowed the river to flow around the dam while it was under construction. Stilled in its course for the first time in millennia, the river pooled, turned back on itself, and rose. Lake Powell was born. It was a triumph for the engineers and the planners and the politicians, but a stinging loss for the conservationists.

No one could have predicted that on that very same January day the Bureau of Reclamation's fortunes would themselves still and begin a slow decline. All the signs looked auspicious. In Washington, the secretary of the interior, Stewart Udall, an Arizonan, had chosen that day to announce the building of more federal dams, including two in the Grand Canyon. The bureau's victory in getting Glen Canyon Dam approved, however, was Pyrrhic. The dam roused so much opposition that it made it impossible to build any more like it. Stiffened by the loss of Glen Canyon, the conservation movement grew adamant in its resistance to any more giant dams in scenic canyons. When Arizona senator Barry Goldwater ran for president in 1964, he announced that he was opposed to the construction of one of the Grand Canyon dams, a first for an Arizona politician. It was probably not even necessary for David Brower to issue his famous Sierra Club newspaper ad, in August of 1966, that asked "Should We Also Flood the Sistine Chapel So Tourists Can Get Nearer the Ceiling?" When President Lyndon Johnson signed a bill in 1968 authorizing the spending of $1.3 billion on new water projects in the Southwest, it specifically forbade the construction of any dam in the Grand Canyon. In 1969 Congress went further and passed the National Environmental Policy Act, which required analysis of environmental impacts prior to the construction of federal projects and likely would have made Glen Canyon Dam impossible. Brower had won.

Nevertheless, Lake Powell kept rising, to his unending regret. The still blue water filled Music Temple and Mystery Canyon and, inexorably, Cathedral in the Desert. It pooled upstream like the incarnation of blind progress: fifty miles, a hundred, a hundred and fifty. It was a hundred feet deep, two hundred, three hundred. It crept up side canyons and stilled more than twenty miles of the Escalante River, with its wildly curvaceous cliffs and alcoves, and more than seventy miles of the San Juan. It seeped into ancient cliff dwellings and flooded the nests the great blue herons had occupied, for as long as anyone could remember, in a grove of oak trees just below Hall's Crossing. The herons tried to nest on the cliff above the drowned trees, and then they vanished. The reservoir lapped toward the feet of Rainbow Bridge. Day after day after day the Colorado, once an outlaw, poured itself into its own dull demise. Some of Brower's friends worried that he would kill himself.

The reservoir, and a huge swath of desert surrounding it, became the Glen Canyon National Recreation Area. To some conservationists it seemed a further slap in the face that the canyon became part of the National Park system only in its demise, and then only in the form of a recreation area that catered almost solely to those who wanted to explore the new waters in powerboats. Before the lake's waters even got there a marina had been built at Wahweap, and the newspapers and travel magazines soon came to laud the exotic new tourist destination: brilliant blue—and highly accessible—water in the most arid and exotic of American places. Soon you could rent a houseboat at Wahweap, and at Bullfrog, and Hall's Crossing, and before too long even at Hite, way up at the top of the lake. You could motor out onto the lake with a barbeque and a fridge and comfortable beds. You could water-ski up winding slot canyons that just happened to be a few hundred feet higher than winding slot canyons you could have hiked up a few years earlier. You could lose yourself in the maze of slickrock. And if you lived in Phoenix you could crank up the air conditioner on a hot summer afternoon and know that the dam operators could reliably provide the juice for you and thousands of your neighbors simply by letting more water drop through the penstocks to spin the mighty turbines.

In 1963 more than ten thousand people visited the nascent recreation area, in 1966 more than three hundred and fifty thousand, and by the end of the century some three million a year. Floyd Dominy, irritated at what he saw as Brower's hogging of the publicity about the place, compiled and printed, at government expense, a glossy color booklet about the new lake. It was titled *Lake Powell: Jewel of the Colorado*. Jewel? Why, yes; any boater could see that it was—a sapphire in the desert!

Or, you could, as many earlier visitors did, mourn. In much less time than it took the Colorado to fill it Glen Canyon became, to those of a certain mind-set, a martyr, an icon, a ready code phrase for blind progress and greed. Even Barry Goldwater, who'd supported the dam, said later that it was the one senatorial vote he regretted. What occupied the canyon now, in much the same way that an enemy army might occupy your land, became known as Lake Foul or Blue Death or simply "the reservoir."

᠀

ON NOVEMBER 21, 2004, I stood next to the four great outlet pipes next to the base of the dam, just downstream from the turbine house and the lawn hidden behind it, and watched as bits of the reservoir shot from the pipes as if from a great fire hose. It was cold and dank in the early morning. Water glistened on every surface: on the sandstone canyon walls, on the pavement of the small parking lot, on the metal walkway that led downstream along the bank. The Bureau of Reclamation had invited reporters and photographers to witness the beginning of the Grand Canyon's second "controlled flood."

The first one had taken place in 1996, after years of planning. By thirty years after the dam's completion, it was abundantly clear that the dam's effects reached not just far upstream, but far downstream as well. The water released by the dam is something a trout could love, cold and clear, and, in fact, the Arizona Game and Fish Department, in the 1960s, was only too happy to create a blue-ribbon fishery for introduced brown and rainbow trout in the reach between the dam and Lee's Ferry, fifteen miles downstream. The anglers wear waders; the water, at fifty degrees, is too cold to stand in otherwise.

The cold water and the new fish were murder on the river's native fish, which were adapted to warm, muddy water and to turbulent conditions. They have inelegant names, such as humpback chub and Colorado pikeminnow. Millions of years of evolution shaped these fish to thrive in an environment of extremes. Their streamlined bodies enable them to negotiate swift currents. Their tough skins protect them from abrading silt. Of the several dozen fish species that swam in the Colorado River basin when the first American settlers arrived scarcely any existed anywhere else on Earth. It was, said Arizona fish biologist Wendell Minckley, "the most distinctive ichthyofauna in North America."

Most of those species had trouble surviving in the still waters of the river's new reservoirs, which were soon stocked with a host of species popular with anglers that often had voracious appetites for native fish: largemouth bass, striped bass, walleye, crappies. Today it is almost unheard of to hook a native fish in Lake Powell or Lake Mead, and there is no chance at all there to catch a specimen of the four biggest native species: the Colorado pikeminnow, humpback chub, bonytail chub, and razorback sucker. The river's ecosystem has become almost unrecognizably different. It's as if the ponderosa pine forest of the Mogollon Rim had all been replaced, in the span of a few decades, by birch trees, except that hardly anyone notices the change.

What's more visible is the way the very topography of the Grand Canyon changed after the dam was built. Those who'd run the Colorado before it may have been more outraged about the loss of Glen Canyon than anyone else, but the dam's construction did make it much easier to run the Grand Canyon. It corralled the high, dangerous flows of spring and early summer and parceled the accumulated waters out in the fall, when the uncontrolled river had sometimes been a mere trickle. It made river running far more predictable—and profitable. Around the time of the dam's construction about two hundred people a year were running the Grand Canyon. In 1970 more than sixteen thousand did.

The dam, though, presented a different challenge to boaters. Dam managers parceled out water releases to produce electricity when it was needed, causing dramatic swings in the level of the river. Rafters might camp on a beach and awaken to find a rising tide carving away at the sandbar they slept on, or they might find their vessels stranded high and dry in the morning, the river having shrunk during the night. Worse yet, the cool, clear water steadily eroded the good camping beaches. With all that sediment getting trapped above the dam, there was no feedstock coming through the canyon to replenish the campsites.

It was to stem these losses that the Bureau of Reclamation conducted the river's first artificial flood in 1996. For seven days the outlet pipes that carry water around the hydroelectric plant ran full. The high water scoured the canyon and deposited sand high up on the beaches. Campsites expanded, as did the still backwaters where juvenile humpback chub like to spend their time, but the good results didn't last. The new sandbars and backwaters quickly eroded. Clear water is hungry water, hydrologists like to say. As if muddiness were its birthright, the crystal discharge from Glen Canyon Dam is eager to snatch sand and silt from the canyon banks.

The flood I watched was a refinement of the earlier one. Heavy rains in the two months previous had flooded the Paria River, which joins the Colorado at Lee's Ferry, and flushed an estimated eight hundred thousand tons of sediment into the main channel. The flood was designed to deposit that material on the beaches and in the eddies. For sixty hours the outlet tubes ran almost full, turning the ordinarily placid waters below the dam into a maelstrom of cold waves and boils. The water represented more than a million dollars in lost electricity revenues, but it was scarcely worth noticing compared with the epic predam floods of one hundred thousand or two hundred thousand cubic feet per second.

One of the men in hardhats standing below the dam, watching the torrent, was Nick Melcher, the bureau chief for the U.S. Geological Survey in Tucson. "Ninety-three percent of the river's sediment is up above that," he said, pointing to the dam. "We're trying to make the best of what's left over. We have to manage with the tools we have, but we're definitely dealing with a diminished supply."

Seven percent doesn't sound like a lot, but any Grand Canyon fish biologist would think it extravagant to work with that percentage of the original resources. So would ecologists in Mexico, who on the old Colorado River delta are trying to squeeze a remnant bit of biodiversity out of a once-fecund area that now receives only a tiny fraction of the water and sediment-borne nutrients it once did. Both groups are working with perhaps 1 percent, or less. Most of the canyon's native fish are entirely gone; the only large species left there is the humpback chub, which barely hangs on. A few months after the flood, however, biologists were finding 63 percent fewer juvenile humpback chub than beforehand. It wasn't clear at all whether there was any way to improve habitat for native fish by manipulating the dam's releases.

Had the flood worked? Well, no one could quite say. "Playing God," an assistant secretary in the Department of the Interior told the *New York Times*, "is harder than it looks." A dam, it turns out, is a mighty blunt tool for managing a complex ecosystem.

⁊

THE *Moon-Eyed Horse* WAS NOT A NEW VESSEL. Its owner had recently donated it to the Glen Canyon Institute for use in tours of the canyon's side drainages now that the water levels had dropped so low. The GCI's young executive director, Chris Peterson (no relation to Andy), named the boat after an Ed Abbey story. In the story Abbey is a ranger at Utah's Arches National Park who tries to capture a renegade gelding that, eschewing all the comforts of ranch life, has escaped captivity and lives by itself in the canyon country, but to no avail, of course. Abbey has as much chance of harnessing the beast as he has of capturing in a canteen the heat and wildness and indifference of the desert—or as Jay Gatsby has of snaring Daisy, for that matter.

The six of us had signed on to the *Moon-Eyed Horse* for a trip that was equal parts sightseeing and service. We were to visit the Cathedral and other scenic side canyons, but we were also to earn our keep by picking up garbage on the reservoir's expanding banks. Tires, beer cans, fishing lures, plastic cups—anything that had been cast overboard in the last forty years might lie exposed on the mud banks now.

Peterson had brought a lot of garbage bags.

Why had they come, the retired schoolteachers, driving many hundreds of miles to spend a cold and windy week picking up litter on a rinky-dink little boat puttering along the muddy banks of a declining reservoir? To Kate Bauer, from northern California, it represented an opportunity she'd never thought she'd have. She'd seen Glen Canyon before the dam was built when she came for a visit with her family in the late 1950s, when she was a girl.

"When that dam started going in," she said, "my parents were really riled up. Right there the whole family decided to never spend a cent on anything the reservoir generated. I never thought it would get this low again in my lifetime.

"What I keep hoping is that this is going to be the next Mono

Lake," she said, referring to the California desert lake whose draining by Los Angeles had been halted in the 1990s. "No one thought
that would be saved, and it was just another little grassroots group
that did it." She hoped that the GCI would be the little David of a
grassroots group that could take out the Goliath of the reservoir.

To see even little bits of the old Glen Canyon would be a treat, all
the women agreed; to pick up the litter along the reservoir banks was
a means of feeling positively engaged in its restoration. No, picking
up litter was not akin to blowing up the dam or even to the political
work of trying to drain the reservoir. It was modest work, but at least
it was something, a way to occupy our hands while our eyes took in
the scenery.

Within a day of starting our trip at Bullfrog Marina we'd established a routine: motor along the lake for a while, stop at a beach or
side canyon for a hike, pick up any litter we saw. We found cans, bottles, fishing line, golf balls, and—the greatest prize—the ubiquitous
white molded plastic chairs that come with every rental houseboat.
Along the way, we would gaze up at the high red-rock cliffs that
formed the top of the old river canyon. Their tops undulated unpre-
dictably, but their midriffs were sharply sectioned by the monoto-
nous white band of the bathtub ring, as mercilessly horizontal as if
drawn by an engineer, as it in effect had been. Wherever we went the
browned carcasses of dead tumbleweeds drifted by us in the light
chop, some as big as washing machines.

What we had signed on to do was a very small echo of the Glen
Canyon Institute's raison d'être—the reversing of what its founders
viewed as a mistake of grand proportions. The institute was founded
in 1995 by a mild-mannered Salt Lake City physician and Mormon
named Richard Ingebretsen. He'd first visited Glen Canyon when he
was a Boy Scout. The reservoir was filling, but getting to Rainbow
Bridge had still required a lengthy hike. His scoutmaster had pointed
to the top of a prodigious cliff. "One day water will flow over the
top of that," he'd said.

"That didn't ring good in my little heart," Ingebretsen told me. "I
remember watching little animals and thinking, 'They're going to
die.' It bothered me. It bothered me."

Ingebretsen became a river runner. He loved to run Cataract

Canyon, just above Glen Canyon. Its lower rapids had been inundated by the reservoir when Lake Powell rose to its highest point in the early 1980s. That bothered him too. By the early 1990s, when he was established in his career, he decided to do something about it. He set out to form an organization that would celebrate Glen Canyon. He set thirteen goals, but restoring the canyon was not among them. He wanted to set up a library dedicated to Glen Canyon; he wanted to collect historic photos of the place; he wanted to make videos. One of the goals was the facilitation of a live debate between David Brower and Floyd Dominy. In 1995 he did just that in a packed auditorium in Salt Lake City. The two grizzled warriors, both long retired, both amiable, picked up right where they'd left off. Brower still considered the damming of Glen Canyon the greatest loss of his life. Perhaps, he suggested, it was not too late. Couldn't the diversion tunnels—the great holes bored through sandstone through which the river had been diverted during the dam's construction—be reopened? Edward Abbey had suggested just that years earlier. Then the river could flow and Glen Canyon could recover, and it would not be necessary to go through the hugely complicated and expensive procedure of unbuilding the dam.

The next year Ingebretsen showed some of Brower's film footage of Glen Canyon, taken in 1962, at a GCI meeting. The board members vigorously discussed how one might go about decommissioning one of the country's major dams. Brower was there. Two weeks later he went to the Sierra Club board. Although he'd been fired as the group's executive director in 1969, he was back on as a respected board member, an eminence who, although grayed, was far from an emeritus.

So it was that on November 16, 1996, the Sierra Club—which by now had over a half million members and a much higher profile than four decades earlier—was suddenly and unanimously proposing to undo a mistake for which it viewed itself partly responsible. Glen Canyon was again national news—but this time it was its unburial, rather than its burial, that made headlines. Politicians ridiculed the idea. Congressman James Hansen of Utah called a hearing to shine a spotlight on what he viewed as the wackiest idea to come out of the environmental community in a long time. It was "the silliest proposal discussed in the 105th Congress." It was "a certifiably nutty idea." It was "nonsense."

Ingebretsen didn't care about the criticism. He is a doctor, and the idea's can-do spirit made sense to him. "In medicine, you have a problem, and you try to solve it," he said. "With Lake Powell, it was, to me, a matter of correcting a problem. It's filling with sediment. It's losing a million acre-feet a year of water through evaporation and seepage into the banks. The Grand Canyon is dying. So let's correct the problem. A lot of people have the desire to fight for something, and Glen Canyon became mine."

He was, he admitted, buoyed by ignorance. He'd never heard of the Colorado River Compact, and he didn't know the central role Glen Canyon Dam occupied in the politics of western water. He learned quickly, though, and cast his net wide. On a visit east he had dinner with Floyd Dominy in Virginia. Dominy was gracious. He apologized to Ingebretsen for the damage Glen Canyon Dam had done to the Grand Canyon. We didn't know that was going to happen, he said. He did not think the controlled floods were going to work.

What about Glen Canyon? What did Dominy think, so many years later, about inundating that place? Ingebretsen wondered.

Dominy slid his glasses down his nose and looked straight at Ingebretsen. "I took absolute *delight* in flooding Glen Canyon," he said. That same evening, though, the engineer in him overrode the politician. Dominy drew on a cocktail napkin a diagram showing how one might, were one so inclined, decommission Glen Canyon Dam. Brower's idea of reopening the diversion tunnels was a cockamamie notion, the engineer said. They were plugged with three hundred solid feet of reinforced concrete. It would be much better to drill new passageways through the softer walls of Navajo sandstone. He sketched it out. "No one has done this before," he said, "but it'll work." "No one will believe you did this," Ingebretsen said. So Dominy signed and dated the napkin and gave it to him. Ingebretsen keeps it in a safe-deposit box.

Still, Ingebretsen's crusade looked mighty quixotic. It was, after all, going up against a lake-based tourism industry that attracts millions of visitors and results in the spending of hundreds of millions of dollars each year, with more than that spent on other business in Page and other towns. It was going up against the renewable electricity generated by the dam—worth something like $100 million a year—at a time when environmentalists were unlikely to smile upon alternative sources such as coal, natural gas, or nuclear energy. It was, above all, going against an entrenched philosophy that says Americans build things; they don't unbuild them.

Then nature weighed in. In 1999 Lake Powell was almost full, but soon drought set in. In 2000 inflow into the reservoir was 62 percent of average. The next year brought about the same amount of precipitation, and 2002 was far worse. On the Mogollon Rim the Rodeo-Chediski fire raged. The Colorado River was down to a quarter of its average volume. It was the lowest flow in recorded history and, according to tree-ring studies, one of the driest spells in five hundred years. By 2003 the surface of the reservoir was down about ninety feet from its full pool level. That spring Ingebretsen was able to boat into the Cathedral chamber and spot the top of the old waterfall a few feet below the surface. It was so close. Maybe, he thought, the drought would continue. Maybe more would yet be revealed.

It was. In 2004 inflow was only half of average, and despite wet weather in the fall of 2004, the reservoir had fallen another fifty feet by early 2005. Ingebretsen went back. This time the water had withdrawn entirely from the Cathedral, although the chamber's basin was filled with far more sand than before the inundation. Ingebretsen was elated. This should not be filled again, he thought; it should be preserved as it is.

But the drought had finally broken, at least for a while. Record-breaking snowpacks lay atop the mountains of southern Utah and southwest Colorado that winter, waiting to melt in the spring sun. The Bureau of Reclamation was predicting that the runoff would raise the level of Lake Powell forty or fifty feet within a few months.

The drought had given the GCI new credibility. The ultimate goal was still to decommission the dam and have the river flow past it unimpeded, but there were more immediate goals, including maintaining a low reservoir level so that the Cathedral and other sights could remain exposed and could begin to restore themselves. Nature, Ingebretsen and his colleagues believed, was on their side. The reservoir was very low—only a third full. So was Lake Mead, which constitutes the main water supply for the lower basin. Surely the Southwest's water could be stored there. Just a month after his first visit to the Cathedral in the spring of 2005 Ingebretsen appeared on ABC's *Nightline*. "We have to store water, there's no question," he said, "but you don't have to store it in our beautiful places and our beautiful canyons."

Nonsense, countered secretary of the interior Gale Norton. She allowed that Glen Canyon had been beautiful and that the dam would perhaps not be built today. Nonetheless, she said, "we have only been able to withstand five years of very severe drought in the West because we had water stored in Lake Powell and Lake Mead."

Simply that the debate had made it to *Nightline* was evidence that the idea of restoring Glen Canyon had come a long way. Whether Lake Powell can ever refill itself is indeed an open question. It took seventeen years to fill the reservoir after the dam was built, and today there are far more demands on the Colorado's water than there were back then. There simply isn't much surplus anymore.

"Lake Powell in the future is going to be mostly empty most of the time," GCI's Chris Peterson said in early 2005. "The era of excess, of water surplus, is over. As fast as water is going to come in it's going to be used. Lake Powell will not be necessary for storage; Lake Mead will be sufficient. At current levels of demand, Lake Powell would be half full half of the time. With increased demand, 60 percent of the time it will be only 20 percent full. You would have to have back-to-back one-hundred-year floods to fill Lake Powell. We are moving beyond talk of draining Lake Powell, because *we* don't have to."

"Lake Powell," Ingebretsen wrote, "has no value or use in the future of water in the American Southwest. Glen Canyon is the future."

Ingebretsen and Peterson's rhetoric might be dismissed as hyperbole were it not for a slew of scientific studies buttressing it. In 1995, before the drought began, hydrologists released a study showing that the Colorado's water supply system would be severely stressed during a prolonged drought. At the turn of the twenty-first century climate researchers analyzed long-term precipitation trends for the region and suggested that the Southwest might be entering a two- to three-decade period of drought. In 2004 a team of climate scientists published results relating much the same story—and adding a new twist. Due to human-induced climate change, they wrote, the Colorado will likely not be able to meet all the demands placed on it. Reservoir levels will be down by more than one-third. Water releases will decline. With lower water levels, the ability to generate hydropower will also suffer.

Given all that, Ingebretsen asked me on the phone, wouldn't it make sense to get rid of what amounted to a giant evaporation basin sitting and off-gassing in the hot desert sun? Lake Powell was said to lose over a half-million acre-feet of water per year just to evaporation. It was not, he said, just a romantic idea of going back to a better time. Getting rid of Lake Powell, he suggested, was a matter of civic responsibility.

⁂

SHE WOULD NOT GO BACK to Cathedral of the Desert, she said. No, no matter how low the water dropped. It upset her to hear people talk of their visits there. The glow of the rock could not be there anymore, she said, not with that mineral stain on the walls. The verdure of the old days was also missing. She had not seen these things with her own eyes, but she knew.

"It was magical," she said. "And I know they say it has come back but it has not, because it can't. It's funny. I don't know why it angers me so"—here she laughed just a little—"but it does, that people are saying, 'It's back. See what can happen; it's back.' It's not back!"

There are two places in Glen Canyon to which Joan Nevills-Staveley will not return: Cathedral in the Desert and Music Temple, which she first saw when she was a girl. Her father, Norm Nevills, was the first commercial outfitter on the Colorado River, and that summer, in 1947, he led a party down the San Juan River and onto the Colorado to Lee's Ferry. She recalled how one of the passengers, a writer named Wallace Stegner who was researching a biography of John Wesley Powell, had written a ditty for her: "There once was a girl named Joanie, who very much wanted a pony." She remembered that she was a handful for the adults on the trip. And she remembered Music Temple. If the Cathedral was the place in Glen

Canyon that looked most holy, the Temple was the place that *sounded* most sacred. Some of Powell's men had carved their names in its soft sandstone.

"The grotto was full of maidenhair fern. It was growing on the slope, almost like a talus, but not rocky," Nevills-Staveley said. "I rolled down that. I was like a cat in catnip. Then Daddy called everybody around to sign the register. We saw the names of Powell's party in the rock. It's been sixty years ago and I still remember it vividly. That tells you it was pretty special.

"I've never gone into that bay. I won't do that."

We were talking at Nevills-Staveley's comfortable, modern house on the edge of Page. Gracious and high-spirited, she was wearing a pink blouse and wire-rimmed spectacles the color of copper. Her gray hair was cropped close. She laughed a lot. It was easy to imagine her, as a young girl, as a handful. Just the week before she'd worked her last day as director of the Page Chamber of Commerce, a position in which she was often called upon to defend the reservoir that had drowned two of her favorite places. She'd worked for the chamber for twenty-eight years.

Nevills-Staveley grew up in the tiny town that today is known as Mexican Hat, Utah, which sits across from the Navajo reservation on the north bank of the San Juan River. In those days it was scarcely a town at all. The population numbered five, which included Norm and Doris Nevills, Joan and her sister, Sandra, and Norm's mother, Mo. In good weather it took two and a half hours to drive the twenty-seven miles to the nearest town to pick up the mail. In bad weather you didn't go.

The family ran a lodge and trading post that served a small and intermittent stream of visitors, but what meant more to Norm Nevills was introducing scores of people to the San Juan and Colorado rivers. He guided a young Arizona businessman and avocational photographer named Barry Goldwater on a sixty-four-day-long expedition from Green River, Wyoming, to Lake Mead in the summer of 1940. He gave Ernie Pyle a good dunking in a rapid on the San Juan. He taught Joan how to row, although she said she wasn't very good at it until adulthood. Joan married a river runner named Gay Staveley. Together, the two of them ran tours of the Colorado, first from Mexican Hat, then from the tiny new marina named Hite at the northeast end of Glen Canyon, and finally from Flagstaff. When the plans for Glen Canyon Dam came out in the mid-1950s she and her husband vigorously opposed them, as did many others who knew the place. It just wasn't right. The engineers and politicians just didn't know what they were doing.

Then she found an accommodation.

"We were devastated when David Brower wrote it off," she said. "So then we changed our tactics. It was really evident there was going to be a dam; there was going to have to be access to the lake. Our business was going to be gone—why not move out to the lake? So, to that end, we began working toward a concession."

The couple applied for and got the government concession to operate the new marina at Hite. They ran trips from there as the reservoir was filling. At times of high runoff the water level would rise a foot a day, and they had to choose campsites high enough so that they wouldn't be inundated overnight. They'd motor downriver through riffles and return a few days later to find them stilled under new slack water.

The new lake was fun: it was a sort of watery elevator that gave entry to parts of Glen Canyon once entirely inaccessible. She was able to find out what was up top of Mystery Canyon and Hidden Passage, and she was able to explore some of the side canyons of the Escalante River drainage, which before had been a long hike from the river.

Over time, she saw a change in the people who came. There were, for starters, far more of them. They didn't come in rafts or rowboats, but in speedboats and cabin cruisers and, soon, in large houseboats. They didn't come solo or in couples or small families; they came in groups and rented a big boat or two. It was different. Gone were the quiet camps and the huge driftwood bonfires and the groves of willow and tamarisk and the pungent smell of the river as it rose in flood and the sound of it purling by at night. But she did not look back, not too much. She liked to see people enjoying the canyons whether they were filled with water or not.

"My theory of life is that when something's over, keep the good memories, but go on. Or look for potential in something new," she said. Her gorge rose when she heard people proposing to drain the reservoir. She thought Richard Ingebretsen and his allies were selfish.

"It's a fait accompli," she said. "The dam is here. It's doing its job. During this drought it has proven itself as a storage basin. To me it's very sad that people should try to focus on something that destroys a lifestyle, if you will. That lifestyle affects millions of people, certainly not just Pageites or people at Bullfrog. Lake Powell is no longer just a lake, it is a home to hundreds of thousands of boaters, and it provides the economy for people in an area that never had one, except for trading posts and things like that. Many people are thinking of this area for retirement. Their money is tied up in their houses. That would ravage those people. I don't think there's been any thought about that. I guess that's been part of my anger, too—or at least I'm quick to rise to the bait. I don't see any real thinking beyond this. You have to think of others as well."

Not much of a lake boater herself, she preferred to explore the backcountry by four-wheeler. Yet she loved the look of the sapphire water against red rock. She liked seeing other people enjoying the place. She was proud of a volunteer cleanup program that began while she was at the Chamber of Commerce: people could ride for a week, free, on a boat named the *Trash Tracker*, picking up litter on shore all morning and having the afternoons free to hike and explore. In 2004 eighty-eight volunteers spent more than three thousand hours collecting almost thirty-nine tons of garbage, including more than fourteen hundred golf balls, more than twenty-eight thousand aluminum cans, five wallets, and close to three hundred chairs. Clearly, the crew of the *Moon-Eyed Horse* was facing some stiff competition.

Sometimes, though, she was all too aware of how many were here together. She had those memories of floating down the San Juan and not seeing anyone for days, except maybe a few Navajo farmers along the bank. She felt a twinge when she thought of how fun and innocent it had been to run the San Juan and of how much her father had loved taking visitors through its rapids. He'd been such a fine, intuitive boatman, so comfortable in that remote setting.

If she had the choice, should the dam have been built? I asked her.

She stared hard out the window at the bright spring sunshine and pink sand.

"That is a hard one," she said, "because I see so much good that has come from it. Very selfishly—*very selfishly*—probably, given the choice, just thinking of me, no one else, I'd say—no.

"Yep. Because I've been very fortunate in being able to see this country almost devoid of people, roads, et cetera, and now, on the other hand, lots of people, and access continuing to increase, and every now and then I get very disturbed because places that I dearly love to go, there are people.

"I said it would be a selfish thing. But you sort of have to think of what the greater good is too."

♣

MAYBE IT IS BECAUSE GLEN CANYON lends itself so little to human occupancy that the past remains so vital there. Unlike Bermuda, say, or Chicago, it is not a landscape that has been worked over, generation by generation, into something that little resembles what it once was. The past is right there in front of you, or underfoot. The cross-bedded lines in a sandstone cliff reveal how the wind blew from a certain direction, millions of years ago, to form dunes that over the ages were solidified into hard stone. Certain caves hold quantities of dung from animals, such as woolly mammoths and ground sloths, that went extinct thousands of years ago; in some cases, paleoecologists say, it is so remarkably well preserved that it smells a bit like fragrant wine. Any side canyon with a bit of flat ground that could be farmed is likely to house, in some south-facing alcove, a cliff dwelling built and abandoned many hundreds of years ago by the people most often referred to as the Anasazi, although many of the choicest ruins were inundated when the reservoir filled. In the mortar of some ruins you can still see, eight centuries later, the fingerprints of the builders. At the aptly named Hole-in-the-Rock Crossing you can see the startlingly rough ravine down which a group of Mormon settlers lowered their wagons in the winter of 1879 to 1880 on their way to settle the new town of Bluff, on the San Juan River. Other cut steps and chiseled holes in the bare slickrock show where miners of the late nineteenth and early twentieth centuries, ever hopeful, tried to blast their way to glory. In Slickrock Canyon, our little group of litter pickers found houseboaters' fire rings a mile inland from the shore just under a cliff house, with its own fire rings, where people had lived eight hundred years earlier. The group may have been cheering against the reservoir, but still there was something a bit melancholy in seeing tamarisk and tumbleweed and cheatgrass growing into the rock circles and covering up the charcoal and beer cans and broken glass. It was as if that sight encapsulated some sort of failure of human ambition.

We picked out what we could but left plenty, I am certain, for archaeologists of the future to find and to argue about.

If all these relics of the past lend a haunting quality to the Glen Canyon area, so do the workings of light and sound. Painting a great east-facing cliff gold at sunrise, sunlight casts a mellow, autumnal glow on the far bank long before its direct rays reach there. People simply *look good* in its light: ruddy, tanned, hardy. Hanging curtains of desert varnish exude a deep burnt umber in direct sunlight and then glisten blue-black like wet and burnished metal in sidelight. Midday sun filters duskily down through narrow stone passageways, spilling from harsh yellows into muted oranges and reds and browns, until it looks as though it had been filtered through the thick woven curtains of some seraglio. The colors are rich and luscious, and often the warm and earthen rock looks as though it were lit from within. Where river or spring water pools in alcoves the sunlight, reflecting off it, dances up and back onto the rock walls, daubing them with dancing curves and lines and arabesques that Katie Lee dubbed "Frank's music," after the friend and boatman—and protégé of Norm Nevills—who guided her on most of her trips through Glen Canyon. It was, she wrote, "a place where you didn't just see color and shadows; the light acted upon you physically. . . . The color was intense, radiant *in* a wall, not merely *on* it."

Textures and topography, too, so unlike those of more heavily settled terrain, can be bewildering in the canyons. Once, hiking down from the rimrock into a side canyon of the Escalante River drainage, I came around a sand dune, looked down into the bottom of the drainage, and stopped. What was a pool of water, shaded by the brilliant green of a midsized cottonwood tree, doing tipped vertical? It took me a moment to realize that I wasn't seeing a pool, of course, but rather a sandstone wall so smoothed by erosion and so painted over with dark and glossy desert varnish that it glimmered just like a still surface of water in the late afternoon light, and the tree's shadow fell in such a way that the mind could read it as either vertical or horizontal. It was like one of those perception tests in which you can see, in a splash of ink, either a beautiful young woman or a chisel-faced old hag—which, come to think of it, is much the way in which people see Lake Powell.

Sound, too, acts in unexpected ways in Glen Canyon. It is a funhouse of echoes. Near the place where I saw a pool turn into a rock wall I stood at the base of a great sand dune while hikers high above neared an ancient ruin tucked into a shallow alcove. I was able to hear every single word they were saying as if they were three rather than three hundred feet away. In the same canyon I slept between a creek and an eroding sandbank and found that if I held my head in one precise position—and there only—the creek very clearly sounded as though it were *in* the sandbank rather than on my other side.

Once, when Katie Lee was hiking alone on a very hot, very still day atop the slickrock domes to the west of what became Bullfrog Marina, she distinctly heard the rock humming like a faraway droning of bagpipes.

"It gets so weird in here sometimes," Lee wrote in her journal back in 1955, "I think I'm *hearing* the light, *smelling* the temperature and *feeling* the sound."

Lee was eighty-five, barefoot, and lithe when I went to visit her. She lives in a funky, sunny old house in Jerome, an old Arizona mining town on the side of a mountain. She starts every morning doing stretches in front of a photo of Glen Canyon; on the good days she is able to imagine herself right back in the place, wading up a shallow-creeked side canyon under a luminous cottonwood. In her sunroom, not far from a sign with the hand-lettered words "The River Always Wins!" is a picture of Lost Eden Canyon—one of the Glen's tributaries—before and after flooding. She'd been telling me about her trips to Glen Canyon after the reservoir had started backing up. She'd been given a boat, named it the *Screwd River*—"with a space between the D and the R," she pointed out—and visited yearly, poking ever farther up the flooding side canyons, until 1967, when she looked down Lost Eden, which had three divinely beautiful forks, and saw that they were flooded with a flat sheet of water.

"I just couldn't go anymore. How to kill a canyon," she muttered as we looked at the photos. "Whew. So, that was it. I kept wondering why I was doing that, because it was so painful. And I guess it was curiosity; when you get so far up a canyon and you can't get any further, you wonder how much else there was there. But the whole thing about it—Jesus! It took me a long time to realize that that's part of the fascination of that place. It's not about to let you know all those secrets. If you're not a bug, or a bird, you're not privileged to that kind of scrutiny. God, isn't it enough what you see? Yes, it is enough! And yet, you know, we would get to the end of those canyons and I was always the first one to try to keep on going. I had to see more. I wanted to see more. You always want to see more. It's just human nature, if you're an adventurous type and you love what you see and you know every time you turn around you're going to see a different new secret of some sort. That's what's so fascinating about it. It took me a long time, after going back over those places that I hadn't been able to get to, to realize that I wasn't *supposed* to get up there. No. Why should I be up here? Everything I saw down below was just utterly fabulous. I didn't have to have the whole thing. It's best to have some mystery hanging over it, and that's what makes it so interesting. There's always a mystery. When the mystery's gone, what is there? If there's no mystery, you might as well go back to the mall. No mystery there."

When a landscape is as disorienting as Glen Canyon, when the sinuous canyon turns rob you of all recollection of north and south, when the bare rock landscape is achingly dry but runs with countless waterfalls after heavy rain, when the play of light is so intoxicating, when the silty river smells of "Mother Earth's perfume," as Lee described it, when the workings of sound and echo rip away accustomed directional markers—then you find yourself opened up to the new, shaken out of your comfortable and unthinking ways of looking at the world. That, surely, is one explanation for why visitors found river trips through the old Glen Canyon so compelling.

꙰

AFTER WE LEFT CATHEDRAL OF THE DESERT, Andy Peterson had a small revelation for us. He had miscalculated the fuel. We couldn't head farther up the Escalante, as we'd hoped. There would be no look at the top of Gregory Natural Bridge—the second largest in the world, after only Rainbow Bridge, and just beginning to show above water—no cruise up to the limit of water navigation, no beguiling glimpses up its winding side canyons.

No one seemed to mind. "If the only thing we saw on the trip was the Cathedral, I'd be happy," one of the retired teachers said. Anyway, we were back out in the sun and warming up. We headed for shore so that we could make camp near the confluence of the Escalante and the Colorado. A couple of us leaped off the bow to fasten the *Moon-Eyed Horse* to what looked like a firm sandbank. It wasn't. We sank calf deep into soft gray muck and had to struggle back out. We heaved a few dozen flat stones into a line to make a primitive landing.

We had, by then, gained an intimate familiarity with the reservoir's mud. We'd seen great pink sand dunes that predated the reservoir in side canyons, their lower flanks worn away by the action of waves during the high-water years. We'd climbed on hardened clay, riven with tamarisk roots, in the canyon bottoms. We'd seen creeks in the side canyons, fed by the wet winter, spilling small fans of fine, red-brown silt into the clear blue or green water of the reservoir. We'd walked up those creeks, where if you joggled a bit you could bring the water welling up into the saturated sand and turn it into a quagmire of jelly that could suck your legs right in. When we walked across that sand we learned how to take quick, light steps so that we wouldn't be pulled in.

But that wasn't as bad as the muck of the still places, the backwater coves and hidden beaches where runoff had spilled from the slickrock. Often what looked like a solid beach revealed itself to be no more than a veneer of sand over a gelatinous and bottomless mass of wet goo. It filled many of the cans and bottles we picked up and weighed down the bow of the *Moon-Eyed Horse*. It stuck to our shoes, smeared our clothes, and soon came to form a sort of a crust on the durable carpeting that covered the boat's deck. Some of it, when wet, wafted a distinct odor of swamp gas—the result of organic matter in decay—or glistened with a sheen of oil spilled from some boat motor.

Still, the teachers liked seeing the mud and the silt. It represented the inevitable end of Lake Powell. "It's fun to see erosion and be glad about it," Bauer said at one point. "It's kind of ironic to be doing Leave No Trace camping when you see what's out here."

More than any argument about aesthetics, more than the fondest, rosiest recollections of what Glen Canyon was like before the dam, more than any disputes about humpback chub or the relative merits of recreating on a lake or on a river, it is the muck that will settle all arguments about Lake Powell. That may not happen in the lifetimes of any of the *Moon-Eyed Horse*'s passengers, but it will happen some day. The Colorado is a liquid freight train carrying load after load of sediment. At flood stage it historically carried enormous amounts of sediment from its arid and highly erodible watershed, over millions of years carving out the Grand Canyon and in some places near its delta laying down sediment more than three miles deep. It has been calculated that the unrestrained Colorado, at its peak flood stage in June, could have filled the Rose Bowl with sand and clay between the beginning and end of a football game. The same amount of sediment still enters the river today, but it settles into the still water of the reservoir. A lot of it lands in Lake Powell: remnants of the gray Wasatch mudstone of Wyoming, of the tumbled glacial moraines of the San Juan Mountains, of the rainbow-colored Morrison shale (rife with dinosaur tracks) that lies about the Uncompahgre Plateau, of the sulfurous minerals poured out of any number of hot springs, of the barren purple Mancos shale and Chinle bentonite, of the great volcanic conduits of the Four Corners region, of the great piled red and orange and cream-colored sandstones of Utah, to say nothing of all the drowned rattlesnakes and coyotes and ground squirrels, of the water-smoothed cottonwood and tamarisk trunks carried off by flash floods, of the uncounted numbers of decaying fish and frogs and mud snails sunk to the bottoms of quiet pools, and of the crumbled log cabins and shattered pots and lost arrowheads and smashed boats and water-logged clothes and, yes, occasional stiffs that form the records of life and death in an unruly watershed. It all piles up in the reservoir. More specifically, it piles up wherever moving water meets still water. Most of it doesn't make it anywhere near the dam. It accumulates, instead, where tributaries of all sizes—whether rivers such as the San Juan and Escalante and Dirty Devil, or tiny slickrock channels a bare hundred yards long that flow only during summer thunderstorms—join the reservoir. The silt of Lake Powell has piled up in endless pockets along its shore, and those pockets move as the reservoir's surface fluctuates. During periods of high water, the sediment accumulates high up in the reservoir. When the water level drops, the rains come and wash the material down to the shrunken pool of still water. It is a process as inexorable and as inevitable as gravity, and even an agency as ingenious as the Bureau of Reclamation has not yet figured out what to do about it. Sedimentation spells the eventual end of every dam. In a landscape as erosive as the Colorado Plateau it simply spells a quicker end than in many more vegetated places.

Various studies have been done on the river's sediment, and the resulting estimates of how long it might take to fill the reservoir have varied from one hundred and fifty to a thousand years. To argue about how long it will take the river to *entirely* fill the reservoir basin with sediment, however, is to miss the point. The real issue is all the fans of debris that pile up wherever tributaries meet still water. Already it has often become difficult for river runners on the San Juan to take out, so clogged with sediment is that river's delta. What was once a thriving marina at Hite is now a wilderness of sand and muck. Many side canyons are difficult of access because of the quagmires that form where the lake water begins. They get flushed out by runoff when the water is low, but then all that sediment simply ends up farther downstream. The real question is not hydrological but sociopolitical: How much mud will people tolerate before they demand that something be done about it?

⁊

THE DAY AFTER OUR VISIT to Cathedral in the Desert we headed back upstream. Our plan was to stop at Hall's Crossing, a marina just across from Bullfrog, fuel up, drop off the garbage, and then motor farther up the reservoir to Moki Canyon, which was said to harbor a nice selection of Anasazi cliff dwellings. For no reason we could discern, the *Moon-Eyed Horse* was listing slightly to star-board. Not to worry—we sat on the port side and chugged valiantly up the lake.

At three miles an hour, we had little to do but admire the tapes-tries of desert varnish that fell all around—and keep sharp eyes peeled for more plastic chairs. We were up to three and wanted to find a fourth undamaged one. Peterson wanted a complete set. Some-where around Annie's Canyon I spotted a white leg sticking out of the sand and tumbleweed wrack in a tiny cove on the starboard beam.

Peterson headed for shore. The *Horse* was moving along a bit sluggishly, and as he shifted it into reverse to reduce speed it wal-lowed. Water washed up onto the foredeck. "Jesus!" he yelled, and pushed forward again. The boat righted itself and we slid up onto the shore without any trouble.

I took off my shoes and stepped off onto the shallow, sandy bot-tom. To my surprise, it was solid. The chair lay upside down, buried in firm muck. Grabbing a paddle, I began to dig. The mud came up brown and black and oily. When I'd excavated a good hole I grabbed hold of the closest leg and pulled.

I heard one of the women say something about Superglue just after that, but as I heard the sharp *crack!* of the leg breaking off I knew that this was one moment that could not be reversed. Instead of a complete set of chairs we now had three whole ones and one irrevocably broken. Some things simply can't be fixed back the way they were.

⅗

ONE OF THE GREAT BONES OF CONTENTION about Glen Canyon is just what happens when the water withdraws. Where the topography is steep, bare rock is left behind; where it is not, sand and silt are left. Beyond that plain observation there is no agreement.

The supporters of the GCI like to point to signs of recovery, such as places where the white crust of the bathtub ring is wearing away, very quickly in some cases, from the red rock. From the deck of the *Moon-Eyed Horse* we did indeed see hanging draperies where the red of the Kayenta sandstone was running down, almost the color of dried blood, over the bleached white rock below it. Elsewhere, the white crust was visibly spalling away from the cliffs.

"Being a climber, I know they've ruined this rock," Andy Peterson said. "The rock's so saturated with water it's really crumbly." The benefit of that, in his eyes, is that the bleached rock deteriorates quickly. The evidence of the rock's wearing away has not stopped him from looking for good climbing locales in the newly exposed cliffs; if he were to find good routes, he knew, rock climbers might become a potent new group lobbying to keep reservoir levels low.

Others, though, can look at much the same scenery and see instead of a promising young red face a dismal white one. "They are forever saying that the canyon will clean itself up," Joan Nevills-Staveley told me. "'A few spring storms and it'll be all cleaned up!' There isn't that much running water in Glen Canyon, even though it sounded like there was. There were canyons that had water, but there were more dry ones than there were wet ones." She thought it would take forever for the white stain to leach out.

A similar argument takes place once vegetation starts to grow. From the boat, after a wet winter, we saw patches of new green beckoning at the entrances of side canyons and on sandy flats perched above the new lake level. They looked promising. But when we got closer we saw that they consisted almost entirely of nonnative plants: thorny tumbleweeds, grasses such as red brome and cheatgrass, and tamarisk—the same small tree David Wingate had planted, and regretted, on Nonsuch Island. It was a bit like seeing the Chicago forest preserves fill in with buckthorn: it looked pleasant and verdant enough from a distance, but once you figured out what used to be there, the new plants could seem more curse than blessing.

John Spence is a botanist who works for the Glen Canyon National Recreation Area. When Chris Peterson, of the GCI, told him that his volunteers were documenting the recovery of vegetation in the side canyons, Spence told him, "'Recovery' is a relative term. The vegetation you're documenting is all exotic."

Spence had by early 2005 seen about fifty or sixty exotic and invasive species of plants growing near the lake: tamarisk, tumbleweeds, Russian olive, rip-gut brome, a new African mustard, tree of heaven. They were virtually all headaches. The tumbleweeds, or Russian thistles, pile up in windrows on the beaches like the thorn fences built in Africa to keep lions out. "They keep the tourists confined on the beaches," he said. The tamarisks suck water from springs and out-compete native willows and cottonwoods. The mustard forms mono-cultures that wipe out native flowers. The thorny Russian olive trees grow into side drainages such as the Escalante and act as giant strain-ers that can puncture rafts. To Spence, more dry land means more room for exotics. The changing water levels, he told me, function as a "continual dynamic feedback loop for weeds, because they're always being wiped out and then always reestablishing themselves. The reser-voir is a giant pump for the seeds of exotic plants."

Spence admitted, though, that some recovery of native species does take place in the riparian corridors of side canyons, where more water is available. I saw that myself when I hiked along the Escalante River in the fall of 2004. The river, seldom more than knee deep, wound its way below prodigious red-rock cliffs and simultaneously through a much smaller canyon made up of sediment left behind by the reservoir. The soft banks reached up five, ten, twenty feet. They were topped with tamarisks and Russian thistles, but here and there were clumps of native willows, or cattails, where it was particularly wet. The banks were visibly crumbling. They slumped and fell and were carried off by the current. It looked as though they wouldn't last long.

Hydrologists and ecologists studying southwestern rivers have found that reservoirs tend to promote the growth of tamarisk. Along a reservoir water levels are either stable or vary relatively slowly over the years. Native trees such as cottonwoods and willows are more likely to thrive along undammed rivers, where water rises and falls very quickly. Their seeds tend to be produced in sync with such nat-ural flood flows and can colonize wet sandbanks quickly after flood-waters recede. The Escalante is a largely undammed river that can flood substantially, so even where it has been inundated it probably still has a high potential for supporting native plants.

Katie Lee had a pithier summation of all that. "Just leave the river alone," she told me, "and she'll rip it to pieces."

Bill Wolverton said much the same thing. Wolverton works for the Glen Canyon National Recreation Area, and he is one of the rare park rangers who still does a lot of ranging. He lives alone in the lit-tle Mormon town of Escalante and spends as much time as he can—whether employment time or spare time—hiking the rugged canyons of the Escalante River and its myriad tributaries. He has had a twenty-five-year-long love affair with the place. I ran into him in Coyote Gulch, which is renowned as one of the most scenic of those side canyons. It was October 2004, and Wolverton was elated because he had just cut down what he believed were the last tamarisks in lower Coyote Gulch.

Wolverton was deeply tanned and was wearing square eyeglasses that would not have been out of place on a Bureau of Reclamation engineer in the 1950s. He was wearing a forest green baseball cap and shorts, a gray Park Service shirt, and ragged neoprene water boots that had logged many miles. He had the stout legs of an invet-erate hiker. John Spence told me that Wolverton could cover rugged terrain at a ferocious pace and managed to nearly kill him, Spence, on every hike the two men took together. I believed it at first glance.

Wolverton was glad to see the recovery of the canyons, no matter how much cleanup work it required of him. He'd first seen Coyote Gulch in 1979, before Lake Powell's water arrived there. Its lower reaches consisted largely of sculpted and scoured rock. Then the reservoir filled. In 1983, when it topped out so high that the bureau's engineers worried that the dam might fail, you could drive a house-boat a few hundred yards up the canyon from its confluence with the Escalante. The water stayed high long enough for many feet of silt to be deposited. As the reservoir fell, tamarisks and willows sprouted in profusion on the newly exposed land. Wolverton took it upon him-self to get rid of the tamarisks. With the help of volunteers, he'd been able to do just that over the course of many visits. It had taken about a dozen years.

"I came down here in '79 and this place was just pristine," he said, pointing to a willow thicket on a ten-foot-high sandbar above a small creek. From the bar's steep and eroded edge, roots groped at the air. "None of this vegetation had grown up in these mudflats. They weren't here. Is the vegetation a bad thing? It is and it isn't. It wasn't here before. But some of it comes back in a heartbeat, like these willows. They're growing where they haven't before. So in some places we're going to get a brand-new Glen Canyon. But there are also places where you can see how much the creeks in these little canyons have washed away. That speaks very loudly to the future of Glen Canyon."

He was excited about Coyote Gulch, and he was excited about the Escalante. "One of the beauties of the Escalante is that flood flows just go right through," he said. "It's still pretty much in its natural state." That didn't mean that it was free of invasive species: there were huge tamarisk thickets along its banks, and the Russian olives kept sprouting. "I'd love to get rid of the tamarisk along it," he said, "but I don't expect that in my lifetime."

Allowing a river to flood is a bit like allowing fire to burn in the forests of the Southwest: it is a letting go, a shrugging off of control—not an abdication, but an acknowledgment that nature is at least as much in control as people are. Along the Escalante, though, the thought of floods coming through and reshaping the vegetation reminded me of Nonsuch Island. To Wolverton, it wasn't a matter of getting rid of the reservoir and letting nature take its course. People would have to stay involved. It would take commitment. For his part, he didn't mind. There was no place he would rather spend his time. He was modest in the pronoun he used to describe his work, but it was clear that it was a labor of his own love.

"This is a really special place to a lot of people," he said, "and we just felt working on the tamarisk was a good idea."

ॐ

WE PUT BACK ON THE WATER. The wind was picking up, and we headed north, into its teeth, as we made for the final bend before Hall's Crossing. We stopped for one more chair that sharp-eyed Kate Bauer spotted on shore. It was whole. That made a complete set.

We rounded the bend, but instead of diminishing the waves increased in size. The lake was wide here. During wetter times it extended far to the north into the shallows of Hall's Creek Bay. The bay had dried up in the drought, but still the wind had a considerable fetch to work with. The waves piled up to a foot, two feet in size. The *Moon-Eyed Horse* bucked and slowed. Peterson turned the boat into the wind again, and spray splashed over its blunt bow. Suddenly, a wave was crashing full force onto the foredeck and rolling through the railing and over our gear onto the carpeting. Someone screamed. "Jesus!" Peterson yelled, and he plunged the boat into reverse. "Everyone to the back!" someone yelled. Had we piled so much garbage on the bow that we were going to founder? Within seconds we'd crowded onto the backseat, a move that had the useful dual effects of tilting the bow out of the water and making it much easier for us to get to the life preservers, which were securely stowed in a rear compartment.

The *Moon-Eyed Horse* wallowed. We strapped the life vests on and clung to one another. Peterson strode to the front of the vessel and began tossing gear, by main force, into the boat's center. We tilted farther to the side. The starboard pontoon was barely holding us above water anymore.

The wind blew hard. The water sparkled. Off the port bow we could see the dark slot, a tantalizing opening into orange slickrock, that was Lost Eden Canyon. All that dry, sun-warmed rock might as well have been the moon for all the good it was doing us now. A thought came to me: in the Abbey story, the renegade horse goes out into the wilderness and *does not return*. Was this, I wondered, really a well-chosen name for a boat?

The *Moon-Eyed Horse*, it seemed, was about to sink in the very reservoir whose existence its owners so decried.

෴

CLIMB A MOUNTAIN, and with each increase in elevation your perceptions encompass more and more. You can see the highways off in the distance, the verdant fields, the smudges of smog and the contrails of jets. Your vision expands, and even though it takes in the grandeur of nature it also, in many places, reminds you just how thoroughly large swaths of this planet's surface have been worked over by human beings.

Descend a canyon and the opposite happens. The markers that point to the human world outside—the airliners, the roads and transmission lines, the orange glow over nighttime cities—drop out of sight, and you are left with nothing but the rock walls, with some water, perhaps, and with some slice of sky above. It could be a sensory deprivation chamber were it not that the sensations the place provides are so powerful. And so your senses are filled, for once, with the wild world. "In that place where no bombardment of unwanted stimuli could enter my sight lines, it was possible to savor the *overview*," Katie Lee wrote of Glen Canyon in her memoir *All My Rivers Are Gone*.

There are plenty of canyons on the Colorado Plateau, but before the dam was built Glen Canyon was a particularly good place to get away from it all. The river canyon and its dendritic complex of side canyons were remote to begin with, and they offered endless opportunities to not only submerge oneself from the marks of humanity, but to become almost subterranean. With all those alcoves and sinuous slot canyons and deeply undercut chambers connected to the world of light above by only a narrow passageway, exploring Glen Canyon was something like spelunking with a helpful splash of sunlight. In no other place were the shape and texture of the rock so outside the ken of how rock usually behaves. In no other place did the rock seem so alive, so sensual, so erotically charged.

In the era after World War II, as many of the West's wide open spaces came to seem a bit smaller, Glen Canyon remained relatively little visited. Visitors stood a good chance of spending days on the river without sighting another boating party. It had been, in fact, more peopled at the turn of the twentieth century than it was six decades later. In the old days you would likely have met gold miners there, sifting the fine sand of the river's bars. You might have run into Cass Hite, a loquacious explorer and miner who settled at a useful low-water ford at the head of Glen Canyon in 1883 and lived there—with a brief stint in prison after he killed a rival miner in a gunfight, in self-defense—until his death in 1914. In the spring of 1901 you might have heard the chugging of a steam-powered dredge designed by an ambitious engineer named Robert Stanton, who, fresh from a failed attempt to build a railroad along the Colorado River from western Colorado to the Gulf of California, spent what probably totaled more than $100,000 of other peoples' money to build a giant sifter that would filter fine particles of gold from the river's sands. Two stories high and more than a hundred feet long, it collected a grand total of $66.95 worth of gold. The company went belly-up. Its president, Julius Stone, came to enjoy river running, and when he returned in 1938 he brewed a pot of coffee on a fire fed by wood from the dredge's wreck. "This cup of coffee," he told his companions, "cost me $5,000." But the view from camp? Priceless. Later in the twentieth century uranium miners did a little better, finding Cold War riches while dumping their spoils right into the river. Ranchers ran their cattle along the river's banks in a few places too. What Katie Lee and Edward Abbey and other river travelers found was hardly a pristine wilderness. There was graffiti: Powell's men and other early explorers had inscribed their names in the soft sandstone, and later visitors emulated them. Below a high, almost freestanding cliff named Sentinel Rock a miner and river runner named Bert Loper had carved his name and all the years in which he'd run Glen Canyon, every few years from 1907 to 1944 (he died in 1949, as every diehard boatman wants to, of a heart attack while running a Grand Canyon rapid). At night Air Force bombers flew low overhead, practicing midair refueling. The groves of tamarisk that lined every sandbar were also a sign of humans—native to the Middle East, the trees were as alien to the southwestern landscape as they were to Nonsuch Island. Yet all these marks were slight in comparison to the overwhelming scale and grandeur of the landscape. They were, in a word, *quaint*, so that even visitors from the Sierra Club, who might have found mining and ranching appalling elsewhere, could view the occasional uranium miner and wandering cowboy more as appealing, wilderness-loving hermits than as capitalists out to rape the landscape. The profit motive simply hadn't been able to win out in Glen Canyon.

Even the most environmentally conscious river runners dumped their garbage right into the river's main current, trusting in its silty, grinding force to turn cans and bottles into harmless new sand in no time at all. They set enormous piles of driftwood—sometimes an acre in size—alight to keep it from rafting downstream and clogging Lake Mead. Norm Nevills lit bonfires and shoved the burning embers down cliffs for the delectation of his guests. In a nation dealing with a Cold War and worldwide responsibilities and the threat of nuclear annihilation, a visit to Glen Canyon was a visit to a simpler time. You could drift down the river, effortlessly, and camp where you liked. You could go naked. You could roll in the mud or in the sand.

You could tan without worrying about skin cancer. Some river runners had friends with small planes air-drop fresh steaks and ice
cream. In 1956 a Utah family built a raft, equipped it with a cow,
pigs, and chickens, and drifted on down the river from Hite. Perhaps
they brought along, for evening reading, a copy of *Huckleberry Finn*,
which includes the line, "You feel mighty free and easy and comfortable on a raft."

"My anxieties have vanished," wrote Abbey of the beginning of
his Glen Canyon river trip, "and I feel instead a sense of cradlelike
security, of achievement and joy, a pleasure almost equivalent to that
first entrance—from the outside—into the neck of the womb.

"We are indeed enjoying a very intimate relation with the river. . . .
Why, we ask ourselves, floating onward in effortless peace deeper
into Eden, why not go on like this forever?"

Why not, indeed? Well, the dam was one reason why not.
Another reason was that—*pace* Faulkner—the past is a country to
which we can't return, however with us it may remain. Those cottonwoods just outside Cathedral of the Desert are dead, and they're
going to stay that way. The dam is not going away anytime soon.
Neither are the tamarisk groves, which were already annoying in
Abbey's day. Even if drought persists for a long, long time it is difficult to imagine the river's native fish thriving at the confluence of
Lost Eden Canyon and the Colorado. The introduced fish that like to
prey on native species are going to stick around whether the reservoir
is there or not. The river's ecological circumstances have simply
changed, in some cases irrevocably.

Recollections of Glen Canyon before the dam are suspiciously
tinged with what is very literally the rosy glow of nostalgia. Everyone
who ran the river back then and is still alive today was young then,
and hale. And a trip through the river canyon could easily be, and
understandably, the highlight of anyone's year or even lifetime. "This
is the way things were when the world was young," reminisced Wallace Stegner, recalling how on his trip you could catch catfish on
hooks baited with nothing more than cigarette butts; "we had better
enjoy them while we can." Indeed we had.

Forty years. A rose red canyon full of sensual rock and water and
mud and rich sunshine. A chance to escape from the rapid pace of

twentieth-century America into a slower tempo. A withdrawal from
the usual responsibilities of earning a living. The effortlessness of
drifting down a peaceful river. Is it any wonder that those who were
there might regard Glen Canyon more fondly than, say, a day at the
office? Of course Glen Canyon is the epitome of a better time and
place. Like Hetch Hetchy Valley, drowned by an earlier generation of
water engineers, it is the equivalent of an athlete dying young, a JFK
of a landscape, preserved in innocence in the collective memory of
both those who were there and those who were not. Were it still
unflooded, Joan Nevills-Staveley pointed out to me, it would be as
popular as the Grand Canyon, and maybe more so. You'd need a
permit to boat there. There would be designated campsites. You'd
have to carry out your wastes, all of them. "All that would probably
be just as hard for the old guard to take as the lake being there," she
said. Katie Lee admitted almost as much in her memoir when she
wrote about her efforts in the 1950s to introduce more people to
Glen Canyon: "I was pooping in my own parlor," she wrote, "introducing a multitude of strangers to my wonderland, where I didn't
really want them at all.

"The ecological Catch-22: Saving a wilderness takes enough people to ultimately ruin it." She also wrote: "Never allow reality to
form a patina over your dreams, or it becomes impossible to reenter
the picture and capture the magic once again."

In other words, it is precisely because it has been so gone that
Glen Canyon has, in recollection, become as much an unsullied paradise as Bermuda was, in anticipation, for the English settlers heading there so many years ago.

✃

IT TOOK A MINUTE OR TWO for us to realize that we were, in fact,
not going to sink. The boat wallowed heavily but didn't list any farther. The starboard pontoon was almost awash, but we could see no
sign that it was going to take us to the bottom. We made for shore,
heading straight north into the teeth of the wind. Peterson kept the
speed down so that waves couldn't creep onto the boat. In five minutes our twin bows were nudging soft silt, and we leapt ashore, heed-
less for once of the muddiness of our landing.

The dispatcher went by the handle of Fast Eddie. "I'll send Captain Tony out," he said, "to save your butts." Bullfrog is a small place; everyone knew that we were the boatload of folks who didn't like the lake. We tied the boat to the shore and unloaded almost everything from it. The garbage we piled just onshore. It was heavy, but removing it did nothing to eliminate the boat's list. There was clearly a leak in the starboard pontoon. The voyage of the *Moon-Eyed Horse* was over, a day early and about two miles short of our goal.

Peterson looked at the heap of garbage, which was considerable. "It's kind of a slap in the face," he said, a crestfallen expression on his face. "We still haven't been able to drop the garbage off. But we won't push limits on this boat any longer."

The passengers sat at the base of a slickrock dune, sun-warmed and out of the wind. Across the embayment we could see a waterspout forming on the wind-whipped surface of the reservoir. My body felt as though it was still moving. I asked the others what they thought of the voyage, and of the reservoir.

"Even if I knew it would end like it did, I'd do it again."

"I was expecting to be overwhelmed by the damage. Instead, I'm overwhelmed by the healing. That's really nice."

"I think this is going to fix itself because it's going to fix itself with silt. But it's not going to do it while anyone who remembers it the way it was is still alive. That means this"—she swept her hand around at the panorama of blue water and red cliff—"might become the norm."

Our conversation was interrupted by the deep whine of motors offshore. A white launch as long as the *Moon-Eyed Horse*, but far more trim, was headed our way. As it neared shore the captain cut the motors and the boat nosed onto the sand.

Captain Tony proved to be a wry young man with sideburns, wire-rimmed spectacles, and a black hat marked "Captain." "How ya doin'? Ya starting to sink?" he called out, jovially. Within five minutes he and Peterson had managed to tie the two boats together, with the *Moon-Eyed Horse*'s sagging starboard pontoon flush against the launch. We loaded our gear and climbed aboard.

"Let's go boatin'!" shouted Captain Tony as he shifted his two big outboards into reverse and began our slow trip back to the marina. He turned out to be one of the few Utahns with an active captain's license. He'd begun working on his when he'd come to Bullfrog to work as a summer employee for Aramark, the company that rents out Lake Powell's houseboats, four years earlier. He liked the work, and the place, and decided he wanted a job piloting what amounts to a houseboat tender. Bullfrog was, he said, "the only place I'd want to live in a trailer." He'd racked up a trove of experience in helping people on the lake. He had helped vacationers whose houseboats had run aground or whose powerboats had sunk because of careless maintenance. He had once, while wearing pajamas, towed to shore a houseboat whose eldest occupant was only thirteen years old after the boat had drifted away from the shore where the parents had stepped off. Since he'd begun working, the lake level had been dropping. The number of people launching their own powerboats at Bullfrog had fallen off steadily, but the houseboat rental business hadn't been much affected. He'd seen a lot of visitors fall in love with houseboating. "Usually they end up loving it," he said, "and they come back again and again." Each year since he'd arrived the boat docks had to be moved farther out toward the middle of Bullfrog Bay, above what had once been a creek.

We progressed slowly in the direction of the snow-covered Henry Mountains, which, as every writer who has ever written about the area has felt obliged to point out, were the last range in the lower forty-eight states to be named. Finally, we reached the boat launch, where Peterson managed with some difficulty to park the *Moon-Eyed Horse* on its trailer. He was already planning to treat Captain Tony to beers and karaoke the next night at the Horny Toad Bar and Grille, the only adult entertainment to speak of in the Bullfrog area. In the meantime the two of them would head back out to pick up our week's haul of trash. Perhaps, Peterson thought, this unplanned collaboration would help bridge the gap between the GCI and the many people who wanted Lake Powell to remain as it was.

When Peterson was done Captain Tony unmoored us from the boat launch so that he could bring us to the marina's dock. Free of the ungainly pontoon boat, he accelerated to a sprightly ten miles an hour. The bow lifted, the wind roared, and the spray flew. After days at three miles an hour, the speed was almost intoxicating.

❧

THE STATES WERE IN LITIGATION FOR YEARS about the Colorado River Compact, so it came as no surprise to anybody that they were unable to agree on what to do about the pleasing surplus of water that fell in the winter of 2004–2005. After years of drought, the winter's heavy snows lay frozen in the mountains like an unexpected inheritance from a long-forgotten aunt. The states squabbled about it as it began to flow. The upper basin states argued that Lake Mead had already gotten a boost because of heavy precipitation in its tributaries; why not, then, leave the upper basin's excess runoff in Lake Powell? The lower basin states argued that the water should run down to Lake Mead because the upper basin states had an obligation to deliver water downstream regardless of precipitation.

The GCI wanted it in Mead also; in fact, Chris Peterson suggested that the reservoir level shouldn't be allowed to rise at all, although he was realistic too. "If I could do anything I'd make it so they couldn't flood Cathedral in the Desert, but that's beyond my law-abiding powers," he said in February of 2005. In the end the states could not agree and so the decision fell to Secretary of the Interior Norton. On May 2 she announced her decision: the status quo. The states had an agreement, she said, and there was no reason not to stick to it. The water would flow down to Lake Mead. Leaving more of it in Lake Powell, after all, would mean a decline in hydropower revenues.

Chris Peterson was pleased. He found himself in the unaccustomed position, for an environmentalist, of applauding Gale Norton. "I think it's fantastic," he said. "It's a step toward making Lake Mead the primary storage facility, and a step toward making water storage more rational. You don't need to store water in Lake Powell."

A few days before he told me that, in early May, he and a couple of friends had taken a raft down the brown, snowmelt-swollen river from just above the old Hite Marina. A sea of sediment between the river and the abandoned marina was outlined by a wrack of weather-grayed driftwood logs and branches that marked the old high-water line. It all looked as easily spilled away as a pile of sand and mud left in a driveway after a really dirty car has been washed. Around the first bend, above the drowned ferry crossing, they'd found a new rapid where the rejuvenated Colorado, although it had been placid here before the dam was built, ran furiously over a bedrock outcropping. They were thrilled; for once, on this heavily used continent, here was virgin terrain, nothing but potential. The rapid was nature, coming back, the river at work. They ran it, paddling hard because their motor wasn't working, and flipped in the cold torrent. Having a river on the move, they learned the hard way, means giving up a certain measure of control.

No one was hurt, but most of the gear was soaked, and a carelessly packed Palm Pilot was lost. Fortunately, Peterson had made arrangements beforehand to have the *Moon-Eyed Horse*, its pontoon repaired, meet the raft crew at the upper end of the reservoir. But the Colorado, whether swift moving or still, was not yet done with the Glen Canyon Institute's fleet. During its trip up the reservoir, the *Horse* encountered a large wave that swept overboard the replacement motor for Peterson's raft.

⁂

BACK WHEN HE WAS IN FLAGSTAFF William Jordan had alerted me to a Robert Frost poem, "West-running Brook," in which a young husband and wife observe a brook that runs west, yet creates standing waves whose crests face up to the east, against the current. It is a poem about restoration, Jordan suggested, or at least about the way people's lives are informed by the past.

> See how the brook
> In that white wave runs counter to itself.
> It is from that in water we were from,

suggests the husband. Existence is, he goes on,

> The universal cataract of death
> That spends to nothingness.

It is a brook, a river, headstrong as gravity. Still, it has that quality of rearing back on itself due to its own turbulence,

> As if regret were in it and were sacred.
> It is,

the husband concludes,

> this backward motion toward the source,

Against the stream, that most we see ourselves in.
The tribute of the current to the source.
It is from this in nature we are from.
It is most us.

Looking back, the poem suggests, is not just what human beings do; it is what differentiates human beings from all the rest of nature. A river moves blindly downslope, unthinkingly. A dam, in stilling its course, tries to stop time. In the end, though, a dam is no more than a card played in a poker game with gravity. What every dam builder knows, but not a one wants to admit, is that the card is always a bluff. The trick is simply to play the hand out as long as possible before you're trumped. The game is rigged. Play long enough, and the house always comes out ahead. The river always wins. A river has its floods that can take a dam out all at once and its muds that can take it out more slowly but just as inexorably. In the long run no dam is going to last. In the long run it is the river, and the gravity that fuels it, that is in charge, in the same way that the mindless passage of time is in charge of us all.

Looking back. If it is the quality that makes us human, it is also a quality that the dam's supporters and its opponents—the Joan Nevills-Staveleys and the Richard Ingebretsens, the Captain Tonys and the Katie Lees—all share in spades. They ought to have a party to celebrate it. Face it, everyone's wistful. Everyone wants to be back in the beautiful rosy past. Things were great when Glen Canyon was pristine and innocent and undammed. Life was sweet when Lake Powell was good and full and no one had to worry about droughts and sediment. Even the hard-nosed engineers were wistful: Harry Gilleland and Floyd Dominy were nostalgic about the dam's construction because it represented a high-water mark of American engineering ingenuity and can-do spirit, the time when the country rolled up its sleeves and got things done, by God. If Glen Canyon was womblike, as Abbey suggested, it is worth remembering that the womb is indeed a sort of original paradise, innocence incarnate, but also a place to which we can't return.

"The past is a sunlit country morning," Howard Mansfield wrote in a beautiful book called *The Same Ax, Twice: Restoration and Renewal in a Throwaway Age*. "The anxiety of today is not present.

Just by being the past it implies a happy ending: See they got through, they succeeded, because, after all, we're here."

We're here, and we're never quite content with the present moment. When a young man named Frederick Dellenbaugh visited Music Temple on John Wesley Powell's second Colorado River expedition, in 1871, he wrote, "This was only one of a hundred such places but we had no time to examine them." Maybe not, but it is the province of youth to hope and expect that such time will yet come. After forty years it is all too clear that it will not. What come instead are memories of a past that in retrospect always appears foreshortened, and for that all the sweeter. In that sense Faulkner was right: the past isn't past at all. We can't go back to it, but it lives on in hearts and hopes. It certainly lives on in the aspirations of those—including pretty much everyone in this book—who look to the past for guidance in how to approach the future.

⚘

THE NEW RAPID WAS ABOUT TO BE DROWNED, albeit briefly, by the resurgent reservoir, but Chris Peterson decided to name it Ingebretsen Falls. By the time his trip was over the reservoir was rising a foot a day. Rivers in western Colorado were flooding. At a hearing in Salt Lake City Bureau of Reclamation officials announced, more or less, that there was about as much chance of decommissioning Glen Canyon Dam as there was of having Lake Powell freeze solid. By the end of summer the reservoir level rose about forty-five feet, and Lake Powell stood about half full, or about half empty. It was a matter of perspective. The floor of Cathedral in the Desert was gone again, at least for a while.

"If you want to see it, come back in October," Peterson told me. "It'll be back." October, it turned out, was overly optimistic. But during his short raft adventure he'd seen how each gallon of water flowing into the reservoir was hungrily chewing away at the gray-green and greasy sediment clogging its old channel, caving in the banks and carrying their substance down a few miles into the slack water, each grain of sand and spoonful of silt further reducing the amount of water that could be stored in the reservoir. He was betting on drought, and on the long term.

For all his hope, though, it's dangerous to grant too much weight to the present. In the 1960s David Brower was suggesting that Lake Powell would perhaps never fill up. It did, more than once. It may be the height of folly to put so much faith in a single drought. The climate models and water-use projections suggest that the reservoir will be largely dry, but to put great credence in them may be to exercise hubris just as the dam builders did. Nature will have its own plans, unknowable to us. Wet years may come. Lake Powell may rise again, or maybe not. Maybe the tamarisk has the upper hand when the water recedes; maybe, in places, the willows and cattails do. Whether you give more credence to one or the other is a matter of perspective and of what time scale you choose.

In the long run, we know for sure that it can't pay to bet against a river and the gravity that feeds it. In the short run, well, sometimes even a tyro can put together a pretty good winning streak.

5

Under the Bridge
of Clouds

On a bright winter day, with the wind whipping hard off the great volcano Haleakala, I hiked up a hill named Moaʻulanui in the middle of the North Pacific Ocean to hear a young man named Derek Kekaulike Mar teach a lesson in responsibility to a group of seventh-grade boys. The boys were attending an alternative school on Maui where they were taught the Hawaiian language and such traditional local skills as ocean canoeing, but otherwise they were pretty much indistinguishable from any other set of eleven-year-old American boys. They were light-skinned and dark, tall and short. Their pants and shirts were irretrievably suffused with red dirt. They squinted into the bright sunlight, punched each other as they hiked, and clutched their baseball caps in the wind.

Mar led us along a rutted red dirt road and to a square, waist-high stone platform about the size of a dining room table. He is a big man, not tall but broad. He is largely of Native Hawaiian ancestry. He pulls his black hair back in a bun and wears wraparound sunglasses much of the time. He is the sort of person who fills a room with the size of his personality, the life of any party. "Derek Mar," he said when we first met, wrapping my hand in his. "You can spell that K-I-N-G."

Mar works for a Hawaiian state agency, the Kahoʻolawe Island Reserve Commission, and he and two colleagues had just led the boys in staking down two long, fibrous erosion-control blankets on the bare cinnamon-colored hillside just below the stone platform. Now he gathered the boys around the platform.

"Now listen up, you guys," he boomed over the wind, and they did. Some of them might have been preoccupied with surfing and Nintendo, but they were about to get a lesson that harked back to more than a thousand years of Hawaiian history. Mar gestured at the platform, which was constructed of water-smoothed stones. Two pointed rocks, one the shape of a shark fin and one mildly phallic, projected above the platform's flat surface. Around them lay a welter of items clearly carried here by human hands not long before: lengths of sugarcane, pieces of white coral, large green leaves tied into bundles the size of a small purse.

"This site is used on a regular basis," Mar said. "It is living. It is called a ko'a. The purpose of this ko'a is to attract rain. Ko'a is the word for coral, and in the genealogical chants of our people the ko'a is the smallest life-form that everything else is built on. Coral reefs attract fish. The rain ko'a does the same thing to rain. There is a certain type of cloud that comes from Maui to here and beyond—the na'ulu cloud. When there was forest on Maui, before the sandalwood harvest and other things, the clouds followed the trees and the trees followed the clouds. When the clouds formed on the slopes of Haleakala they came over here and got caught on the trees of Kaho'olawe. When the winds wrap around they bring the cloud bridge to Kaho'olawe. You hear about it in the old days, in the chants.

"It would be too much to irrigate every single plant here. Mo' betta you call back the rains. In the Hawaiian thinking there's no separation between the physical and the spiritual. The exchange is that you give the gods all these gifts"—here he gestured at the ko'a and the offerings left on and around it—"and they give you back the water."

It had, as it happened, rained on the two days preceding our visit, but you could hardly tell. All around us was a landscape of bare and scarred red dirt. The persistent trade winds had sucked out any moisture that didn't immediately run off into the ocean below. It was as unhealthy-looking a piece of land as any I'd seen during my travels. But what Mar was saying gave reason to hope that this degraded place might, perhaps in the lifetime of the boys who glanced uncertainly around, again be swathed in vegetation that would gentle the rains and stop erosion from carving away at the soil.

It might do much more too.

The idea that the physical circumstances of a place might rely on the beliefs of the people who live and work there, and on their regular demonstration of those beliefs, is a challenging one. Ecology, after all, is a science. The pages of ecological journals are full of measurements and tests of statistical significance. Ecologists count and measure and monitor; they are only too happy to talk about the effects of pollutants or hunting or habitat fragmentation or climate patterns on the organisms they study. From such quantities they can make peer-reviewed mathematical predictions about how a particular species will respond. There's a cause, and there's an effect. Because the natural world is complex, there's no end of argument about the particulars, but every ecologist agrees on the broad outline: something happens in the environment—a fire or flood or clear-cut or new parasite imported by people—and it affects all the plants and animals that live there. Ecological restoration picks up on that understanding and suggests positive actions that people might take to swing environmental conditions in one direction or another: an erosion-control blanket may stop the gullies from expanding and the hillside from washing into the sea, or the planting of seedlings may be the germ of a future forest.

How, though, can you measure what effect the construction and use of a koʻa has on the rain? Can the science of ecology accommodate the idea that the reforestation of an island might rely not only on erosion control and plantings, but also on the practiced beliefs of the people who felt moved to build and visit a stone shrine? Put it this way: could it be those chunks of sugar cane and the other offerings on the koʻa—or more accurately the devotions of the people who put them there—that bring rain, and perhaps someday a forest, back to Kahoʻolawe?

That was the question that brought me to Hawaiʻi. If the forests of the southwestern United States present an opportunity to integrate the restoration of the natural world into the modern economy, Kahoʻolawe, where the American colonial project reached one of its limits, presents an opportunity to integrate it into something much greater: the cultural and spiritual life of a people.

⚘

BACK AT ONE OF THE NORTH BRANCH WORKDAYS outside Chicago I'd mentioned to Laurel Ross of the Field Museum that I was going to write about a restoration project in Hawai'i. She'd recently traveled there to teach local conservationists how the Chicago Wilderness consortium had been organized. Hawai'i was under siege by invasive, nonnative plant and animal species, she said. Especially in the lower elevations, where people live, it was often hard to find a single native plant. The Chicago preserves, with all their buckthorn and garlic mustard, were by comparison a paragon of health. "People think Illinois is bad in terms of what we've lost, but man, Hawai'i—it's pretty daunting," she said. "I'm not sure I could keep my spirits up if I worked there."

It was, in truth, a melancholy time to visit Hawai'i. Just a day or two before my flight out it was announced on the radio news that a *po'ouli* had died. Even National Public Radio doesn't ordinarily trouble itself with the passing of a single, rather drab songbird. This po'ouli, however, was very likely the last one in existence.

Until 1973, scientists knew of the po'ouli only from fossils. That year a group of University of Hawai'i students was conducting fieldwork on the windward slopes of Haleakala, which are cloaked with rugged, wet forests. Amid the misty, tangled trees they spotted a black-masked gray bird they'd never seen before. No one had. Quick surveys of the area revealed that there were fewer than one hundred individuals, so the new species, whose name means "black mask," immediately joined the long list of Hawai'i's imperiled forest birds.

The Hawaiian archipelago, isolated from all other lands by more than two thousand miles of ocean, has among the most unique avifaunas in the world. Dozens of species of land birds evolved there and don't occur anywhere else. Their isolation and the limited extent of their habitat, though, have made them very vulnerable, like the birds of Bermuda, to all sorts of ecological changes. After the first Polynesian settlers reached Hawai'i about seventy bird species are believed to have become extinct; since Captain Cook's arrival, in 1778, at least twenty-four more have, and at least thirty more are in serious trouble. On the entire North American continent north of Mexico, by contrast, only about a half dozen bird species have died out in the last two centuries. Other groups of native Hawaiian

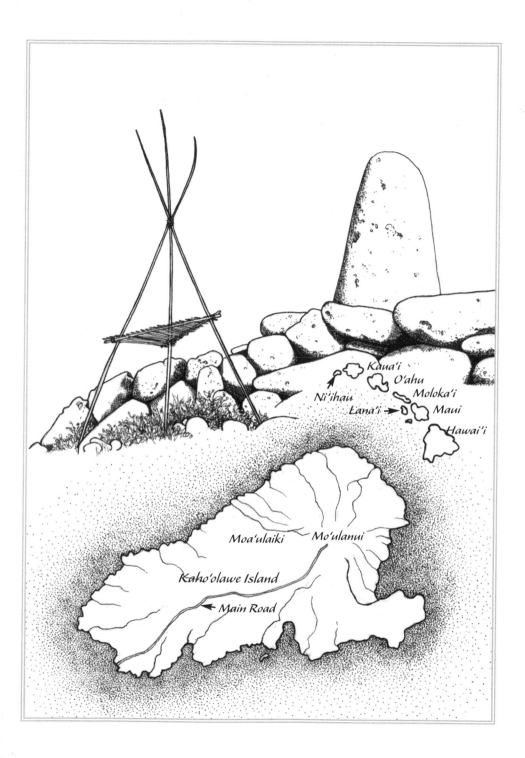

Kaua‘i

O‘ahu

Ni‘ihau

Moloka‘i

Lāna‘i

Maui

Hawai‘i

Moa‘ulaiki

Mo‘ulanui

Kaho‘olawe Island

Main Road

organisms have been hit almost as hard: large numbers of land snails are known to have vanished from the islands, as have 10 percent of the archipelago's plants. Although Hawai'i is tiny in comparison to the mainland United States, a third of the plant species on the federal threatened and endangered species list are Hawaiian.

To hear that the single po'ouli in captivity had died, then, and that a bare two others might or might not still exist in the wild—they hadn't been seen for months—was not really much of a surprise, but it was a severe reminder of what can happen when people alter a fragile place to fit their needs. I went to Hawai'i expecting the worst.

⚜

KAHO'OLAWE SITS IN PLAIN SIGHT to the southwest of Maui, a low mound of land breaching the horizon like the back of one of the many humpback whales that populate the 'Alalakeiki Channel in winter. Forty-five square miles in extent, it slopes gently upward from the shore and, except along its cliff-girt south coast, lacks the precipitous topography so characteristic of the other islands. Like them, it belched up from the depths a million or two years ago amid copious hiccups of lava-cooked steam: there was blue water, then there was hissing black rock, then there were plants, and birds, and languid whales cruising the sun-warmed shallows.

And then there were people. It's believed that the first human beings arrived in Hawai'i, on deliberate voyages from Polynesian islands far to the south, around the time of Christ. Archaeological evidence reveals that by the early eleventh century they were spending time on Kaho'olawe. The island rises to no more than fifteen hundred feet and sits in the rain shadow of Haleakala, whose heights reach up more than ten thousand feet and block moisture-bearing trade winds; it has always been a dry land that couldn't provide the splendid farming opportunities of Hawai'i's larger and lusher islands. But it did offer access to a rich storehouse of aquatic life. Fish and shellfish thrived in its coral shallows; monk seals and sea turtles hauled out on white beaches to bear their young. All around the island's coastline are scattered ko'a, built of rock, that probably pointed the way to productive fishing grounds and provided fishermen a place to leave offerings to the appropriate gods.

The low elevations of all the Hawaiian islands, including all of Kahoʻolawe, have changed to such a degree in the last two hundred years that it is difficult to say with any authority just what they looked like before that. Kahoʻolawe was probably covered with a low, wind-sculpted dryland forest of trees and shrubs and grasses. Low depressions filled with water during winter rains, replenishing underground aquifers. There may not have been any perennial streams, but people were able to collect fresh water from shallow excavations in drainages or hollows.

The island may have been peopled year-round at times and only seasonally during others. It was surely altered by human usage. On other Hawaiian islands it is known that people burned areas to provide space for growing taro, sweet potatoes, yams, and other crops, and perhaps to promote the growth of grasses used for thatching. Kahoʻolawe was not pristine, but it was integrated into the social life of the Hawaiian archipelago. On Maui land was traditionally divided into units called *moku*, large tracts of land that reached from the uplands to the sea and contained a great variety of habitats and resources. Moku were further divided into watershed units called *ahupuaʻa*. The whole of Kahoʻolawe was an ahupuaʻa belonging to one of Maui's moku, so it is likely that the same people used both places.

Kahoʻolawe entered the consciousness of the western world in 1778, when Captain James Cook became the first westerner to safely land in Hawaiʻi. After he was killed on Hawaiʻi's Big Island in February 1779, his ships sailed past Kahoʻolawe's southwest tip and the sailors sighted "neither houses, trees, nor any cultivation." That the island's south coast happens to be bordered by tremendous cliffs, where natives would have had no reason to place their houses, was a fine point not noted by the travel-weary mariners.

In 1793 Captain George Vancouver of the British Navy stopped on Maui, fresh from exploring the island that came to be named for him off Canada's coast, and left a gift of two goats to one of the local chiefs. It seemed a good idea at the time: like Bermuda, the islands of Hawaiʻi evolved in great isolation, without any large mammals. With the exception of the small pigs, dogs, and chickens that the Polynesian settlers had brought with them, the land didn't support food animals.

Chief Kamohomoho of Maui told Vancouver that ongoing wars had left Kaho'olawe and the neighboring island of Lana'i "nearly over-run with weeds, and exhausted of their inhabitants." Goats must have seemed a neat solution. They could survive anywhere, could take care of themselves, and would constitute a ready pantry for both natives and visiting sailors. The two were soon placed on Kaho'olawe.

The goats' success was to prove beyond anyone's expectations. Hawai'i's age-old isolation, and absence of herbivores, meant that its plants lacked defenses such as spines and poisons. The agile goats found virtually everything palatable and began eating plants all across Kaho'olawe down to their nubs. Instead of ameliorating the island's exhaustion, the goats made it far worse. They proliferated wildly.

At any other time the introduction of goats might have been an extraordinary innovation, but against the backdrop of the times they were small potatoes. Hawaiians of the early nineteenth century lived through the greatest changes that had taken place since the first people arrived there. Ever-increasing numbers of American whalers, merchants, and missionaries showed up. They didn't bring only new species of livestock, but also such inadvertent gifts as smallpox, venereal disease, cholera, and mosquitoes. The native population plummeted from at least four hundred thousand—and perhaps much more than that—in 1778 to less than half that by the time King Kamehameha died in 1819. His successors, for political reasons, quickly abolished the ancient *kapu* system of laws and beliefs that had regulated Hawaiian society for generations. That left a void that missionaries were only too happy to fill. A delegation from Boston arrived in 1820, and within a few years Hawai'i's rulers were professed Christians. There were churches and western-style schools; there were shops; there was sugarcane to be grown and money to be made. Within the span of less than a lifetime Hawai'i had gone from an utterly independent land with a natively evolved culture to what the new arrivals, at least, considered an outpost of the new U.S. empire.

Kaho'olawe, though, was a hard nut to crack. There was no good harbor and it was too dry for farming. Only a smattering of fishermen and their families lived there. The missionaries, however, had arrived with new laws, including laws against adultery, which in native culture was regarded as no sin at all. There were, to the missionaries, many sinners in Hawai'i, and many criminals, and Kaho'olawe was turned into a penal colony. It wasn't a very effective prison, though. Several times inmates swam to Maui or Lana'i for food, boats, or women. In 1852 the last prisoner went home. Something else would have to be tried if Kaho'olawe were to contribute to the modern economy.

That something else was ranching. It was evident that the island could support livestock: all those goats were already there. With a gold rush economy under way in California, there was a ready market for fleece and mutton. In 1858 two U.S. officials of Hawai'i leased the island from the government for which they worked. The next year they shipped more than two thousand sheep. The Native Hawaiians living on Kaho'olawe became ranch hands. By 1875 there were twenty thousand sheep on the island, and by 1879 it was reported that Kaho'olawe's upper plains were "entirely denuded of top soil . . . the whole interior plain has been so swept by wind and floods, that nothing but a very hard grit is left."

What happened to Kaho'olawe's trees, shrubs, and grasses echoed what was happening on the other islands. In the early days of American contact so many sandalwood trees were cut down and their fragrant and valuable wood shipped to China that the species was virtually extirpated from Hawai'i. Throughout the nineteenth century vast swaths of forest, including stands of enormous koa trees six feet in diameter, were cleared to make way for cattle ranches. Extensive upland forests were reduced to a few fragments. Instead of soaking into lush vegetation and replenishing streams and aquifers, the heavy rain of the mountains came to run off all at once. To their chagrin, the planters who were making fortunes on sugarcane noticed that water supplies in the lowlands had become irregular and the floods more damaging.

One result was the creation of a territorial forest reserve system in 1903, not long after the islands were annexed by a United States hungry for expansion. Teddy Roosevelt's Progressive agenda, including his protection of the great western forests, was in the air. By 1910 the sheep ranching efforts on Kaho'olawe had failed for good, and Governor Walter Frear of the new territory of Hawai'i declared the entire island a forest reserve. It was an odd sort of forest reserve because it almost entirely lacked trees, but Frear must have hoped that the declaration, and protection from grazing, would cause trees to grow. His specialists told him that the island might present an opportunity to study the connection between tree cover and rain, as it was widely recognized at the time that the naʻulu rains over Kahoʻolawe had ceased with the destruction of its trees and those once on the heights of Haleakala.

The trees, and the rains, weren't coming back on their own, however. In the afternoons the plume of red dust kicked up by the trade winds was visible from Maui. In 1918 Angus MacPhee, who ran a large ranch on Maui, leased the entire island as pasture for cattle and horses. He was from Wyoming, and ambitious, and unfazed by badlands. His lease agreement held that he was to get rid of the goats and stem the erosion. During the first four years his cowboys captured thirteen thousand goats and shipped them to Maui for sale. Those they couldn't catch they shot, leaving only a few dozen of the wariest alive. "The land," MacPhee's daughter wrote in her memoir of the island, "smelled horribly of death."

MacPhee's efforts to revegetate the island were backbreaking. He brought in many thousands of plants, from native trees to such imported species as eucalyptus trees and Australian saltbush shrubs, which still grow there today. He released turkeys, pheasants, and quail for hunting. His family raised watermelons and papayas and eggplants at the ranch headquarters on the island's north shore.

It was a good life. Horses and cattle raised on Kahoʻolawe brought a profit. Parts of the island were turning green. But other areas, especially large parts of the uplands, retained the "very hard grit" that made MacPhee, the transplanted westerner, feel right at home.

৵

"Now," said Paul Higashino, nosing the SUV down a rutted slope that appeared more gully than road, "I'm going to take you to the worst-case scenario. I'm going to take you to the place where NASA really did send the Mars Rover."

Had it not been for the burnished nickel of the ocean surface banding the horizon and for the snowy peaks of the Big Island rising like improbably high mirages through the haze, it would indeed have been easy to suppose we were following *Opportunity* around the red planet, or at least near my home in northern Arizona, among the desiccated badlands of the Painted Desert. The earth was cinnamon red before us and bare, and it stretched in hardpan plains for at least a mile before it dropped off, abruptly, at the top of sea cliffs whose faces we couldn't see. Throughout the barren flats tiny sharp-edged cracks split into rills that widened into gullies that ran down into canyons with sheer walls thirty feet high. Tinted in rust and burnt sienna, mustard yellow and lavender and rich red ocher, the colors of their exposed strata would have made an artist from Santa Fe blush. In some areas boulders the size of bowling balls rested atop tall earthen pedestals whose height hinted at how much of the soil had been carried off by erosion. Scarcely a plant grew. When my eye lit upon a scraggly tuft of buffelgrass, or, off in the distance, a *kiawe* tree, it lingered involuntarily and a bit gladly, as if encountering an oasis in the desert.

Higashino is a soft-spoken, middle-aged ecologist who grew up on Oʻahu. His wrinkled face bespeaks something of the time he's spent working outdoors. He is impassioned about the work he does. He and his wife, who is also a biologist, named one of their young sons after a type of native grass. Before coming to Kahoʻolawe he'd worked at conservation jobs in the lush forests of Maui and the island of Hawaiʻi, eradicating pigs, building fences, taking out invasive plants, cataloguing the native ones. It was against his expectation that he'd come to love this dry and depleted place. As he drove he was explaining how it had happened.

"It was 98 percent bare up here, as far as the eye could see, when I first came up here in 1978," he said. "I thought, 'Why work for an island like that? Why not for some area that would benefit the masses?' Then I started coming out here more and fell in love with it. Even the hardpan is beautiful."

The hardpan is the single biggest ecological challenge on Kahoʻolawe. It is refractory terrain, with the consistency of dried cement. It is almost impossible for water to soak into it or for seeds to become established. If they don't blow away in the persistent winds, their roots lack the strength to penetrate it. Higashino has in the past likened restoring the island's hardpan to revegetating the

Costco parking lot back on Maui. And those gullies keep cutting back farther. When their sediment runs off the island—and surveyors have estimated that almost two million tons of it do so each year—it doesn't just dissipate in the ocean. It accumulates on what once were fertile coral reefs, smothering them.

Higashino once worked at Volcanoes National Park on the island of Hawai'i, which is one of the most geologically active places in the world. The volcanoes Kilauea and Mauna Loa continue to erupt there, coursing their lava into the sea and forming new land. Sometimes the hot rock runs into a forest and converts old stands of trees into a smoking plain of ropy black lava. But quirks of topography allow some stands to remain untouched. The Hawaiian name for such a place is *kipuka*. A kipuka is an island of life in a landscape otherwise entirely barren, a refugium from which lichens and trees and vines can recolonize the new rock. The patches of vegetation on Kaho'olawe's southern plain looked like kipuka to me, so isolated were they amid the sere red dirt.

You might see the all-but-bare hardpan as the end point of devastation. Then again, you might also see it, like a fresh lava flow, as a starting point with nowhere to go but up.

"I came out with the attitude that you can do anything here," Higashino was saying. "It's going to take a while, but it's an opportunity."

৵

ON KAHO'OLAWE, in the twentieth century, the primary battle was not about goats and hardpan; it was about bombs. Even before World War II started the managers of the Kaho'olawe Ranch allowed the Army Air Corps to use the southern part of the island as a bombing range. On December 8, 1941, with smoke still billowing from the wrecked ships and planes at Pearl Harbor, the military declared martial law and appropriated the entire island as a bombing and training range. The ranchers were allowed to get the horses off the island, but the MacPhee family was never compensated for its loss of property and livelihood. The island and a marine zone extending up to two miles offshore were closed off to all nonmilitary access.

In the months that followed, submarines fired torpedoes at targets painted on the cliffs. Naval gun crews worked out the trajectories of artillery shells at the expense of the ranch buildings and water tanks. Landing craft slid up the sandy beaches, preparing Marines for Tarawa, Okinawa, Iwo Jima. Kahoʻolawe was a perfect place to practice on: it was close to the bases in Hawaiʻi, had no resident population, and provided a wide range of targets.

It was so perfect that once the war was won, the Navy didn't want to give it up. Angus MacPhee wanted his ranch back. He thought he might import pigs that, in their rooting, might locate any ordnance that hadn't exploded upon impact. But there was a Cold War to fight, and that nasty little skirmish in Korea. Hawaiʻi wasn't a state yet, and it was a long, long way from Washington, D.C. Local concerns were easily brushed aside. In 1953 President Dwight Eisenhower signed an executive order granting the exclusive use and jurisdiction of Kahoʻolawe to the U.S. Navy. The island, said his order, was to be "restored to a condition reasonably safe for human habitation when it is no longer needed for naval purposes."

For the next four decades the Navy threw everything it had shy of nuclear weapons at Kahoʻolawe. Abandoned vehicles were towed out onto the hardpan and parked as targets in the middle of enormous circles of tires painted white. Planes shrieked in, dropping everything from three-pound practice bomblets to live two-thousand-pound bombs big enough to take out a railroad bridge. Ships pounded the island with small-arms fire and with sixteen-inch shells filled with 150 pounds of high explosive. Sometimes shells hit the ground at an obtuse angle, ricocheted upward, and bounced clear across the island, startling fishermen in the waters to the south. Marines camped on the island practiced firing mortars whose shells were outfitted with antipersonnel explosives or with white phosphorous flares that turned night into day. Torpedoes ripped away at the cliffs and sea stacks. Shrapnel, and the bombs and shells that happened not to explode, filled gullies and littered the hardpan plains. At the lower elevations, where plants still grew, the tenacious roots of kiawe trees and buffelgrass wrapped themselves around rusting fragments of metal and frayed ends of parachute cord that led to who knew what.

From Maui, eight miles away, it looked like the *Star-Spangled Banner* come to life. Bombs burst, rockets glared. The sky lit up and the

windows shook. They never did so more than in 1965, when the
Navy three times blew up igloo-shaped piles containing five hundred
tons of TNT on Kaho'olawe's southwest tip as part of Operation
Sailor Hat, which was intended to simulate the effect of a nuclear
blast on nearby ships. To this day those are believed to have been the
largest nonnuclear explosions ever caused by humans. The ships were
hailed on by rocks, but were not seriously damaged. The blasts did,
however, leave an enormous water-filled crater fifty feet deep and
more than a hundred feet in diameter on the shore. They produced so
much heat that rocks melted, then re-formed themselves with colorful
sherbet swirls as they cooled. From the air the crater looked like a
well-formed grommet, as if the Navy, not content with Kaho'olawe's
location, were preparing to tow the whole island farther out into the
Pacific so that even bigger explosions could be attempted.

To many Hawaiians what was happening on Kaho'olawe was simply the most blatant manifestation of what was happening throughout the islands as a whole. Land was expropriated from Native
Hawaiians, and access was denied to traditional fishing and hunting
areas. Newcomers from the mainland were snapping up real estate
and inflating prices and property taxes. Locals who couldn't afford
to buy even a scrap of land were relegated to jobs as waiters and
cooks and landscapers for wealthy visitors. Streams that had in the
past watered plots of taro and other crops were diverted to huge
pineapple and sugarcane fields. In schools, the few children who still
spoke some Hawaiian were strongly—often corporeally—punished if
they were caught speaking the language.

Hawaiian culture seemed down and out. The daily explosions
over on Kaho'olawe only reinforced how far it had fallen. The island
may have been only a few miles from Maui, but it might have been
Mars for all the immediate contact Hawaiians had with it. Like their
culture, it was battered and bruised, badly in need of rebirth.

๛

THE MODERN STORY OF KAHO'OLAWE really begins on January 3,
1976, when a boat pulled up near the island's shore and nine people
jumped out. This act was illegal, and seven of the nine were almost
immediately arrested by the Coast Guard. Two men, though, hid

among the kiawe trees and were able to remain on the island for a couple of days before they too were arrested for trespassing. Back on Maui, they told of what they'd seen: Kahoʻolawe, although sadly mishandled, was no wasteland. It had beautiful sandy beaches and rocky shorelines and groves of trees. It had archaeological sites. It had been some place. It could and should be again.

That initial trip hit a nerve, and not just in Hawaiʻi. It was time, American Indian Movement activist Russell Means remarked, to "blow out the candles on the white man's two-hundredth birthday cake." Other occupations of Kahoʻolawe followed, as did the formation of an ad hoc group called the Protect Kahoʻolawe Association.

"In the 1960s and 1970s many young Hawaiians went off to college, and when they came back they became aware of the voice they had and how they could use it," said Pualani Kanahele, a hula master and professor of Hawaiian studies at Hilo Community College. "We felt we needed to get some of our land back, and Kahoʻolawe was the most obviously abused place."

In February 1976 group members petitioned the Navy for official permission to visit the island to perform religious ceremonies. Permission was granted, and on February 13 sixty-five Hawaiians landed on Kahoʻolawe to perform ceremonies dedicated to the island's cleansing.

Group members soon filed several lawsuits against the Navy, charging it with numerous violations of laws protecting the environment, cultural resources, and religious freedom. They talked of the need to practice what they called *aloha ʻaina*, a phrase that translates as "love of the land" but that really refers to a reciprocal, even symbiotic, relationship between the health of the land and the health of its people. Kahoʻolawe, they said, was a *Hawaiian* place that, unlike the other islands with their high-rise hotels and waves of tourists, could truly be Hawaiian. It was badly abused, yes, but it could be as much part of Hawaiian culture as the native language, hula dancing, and ocean canoeing, all of which were undergoing revivals at the time.

So that it could not be sued by the Navy, the grassroots association didn't incorporate itself, but it did, under the guidance of an elder named Edith Kanakaʻole, change its name to Protect Kahoʻolawe ʻOhana, or PKO, after the Hawaiian word for "family." Its members, sensing momentum, planned more unauthorized landings. On January 30, 1977, five men landed on Kahoʻolawe. One of them was a gifted guitarist and vocalist, George Helm, who was a rising star in the growing revival of traditional Hawaiian music. He and two of the others gave themselves up, but not before seeing to it that the remaining two activists set up a hidden camp where they could stay.

Helm was eloquent. Two weeks later he spoke to the members of the Hawaiʻi State House of Representatives. "We are motivated to pursue the action of protecting whatever is left of our culture and very basically, it is simple," he said. "The culture exists only if the life of the land is perpetuated in righteousness. . . . The bombs over there, for me, is not the danger; it's the negligence." The House of Representatives passed a resolution condemning the ongoing bombing. To lesser effect Helm wrote to President Jimmy Carter, petitioning for a halt to military activities: "The island is 45 square miles, potent with life—it is not a barren rock and the barrenness is in those souls that see it as such."

A few days later the PKO planned a mass invasion of the island, but called it off when the Coast Guard was tipped off. There was word that boats making the crossing over to Kahoʻolawe might be confiscated. Helm was worried. No one had heard from the two men who'd begun occupying the island in late January. Early on the morning of March 6 he and two companions, Kimo Mitchell and Billy Mitchell (the two were not related), were ferried to the north side of Kahoʻolawe to look for them. The threesome made it ashore by paddling two surfboards. They couldn't find any sign of the two men on the island who had, unbeknownst to the new arrivals, just turned themselves in to the Coast Guard.

Helm and his companions were supposed to be picked up the next night, but the boat, its drain plugs removed, had sunk in a harbor on Maui. When the boat didn't show, the three men took to the surfboards and tried to make it back to Maui. The surf was high. Billy Mitchell eventually turned back to Kahoʻolawe and was picked up by a Coast Guard patrol a day later. Helm and Kimo Mitchell were never seen again. A few days later their surfboard turned up drifting in the Kealaikahiki Channel far to the west of Kahoʻolawe. Helm and Mitchell became martyrs. To the conspiracy minded, the deaths were no accident. Why had the relief boat been sunk? Why had the three supposedly left Kahoʻolawe in the high surf? Helm was charismatic, a born leader. Many thought he'd been murdered.

After the disappearance of Helm and Kimo Mitchell the campaign to stop the bombing focused less on civil disobedience and more on the courts. In September 1977 a federal judge ordered the Navy to inventory and protect important cultural sites on the island and to prepare an environmental impact statement. A year later the state and the Navy signed an accord agreeing that they would work together to eradicate the goats, revegetate the island, and protect archaeological sites. In 1980 another court settlement required the Navy to limit the areal extent of bombing and begin clearing unexploded ordnance from Kahoʻolawe; in addition, it was required to allow the PKO regular monthly access to the island.

The campaign to return the island to Hawaiian control took another decade. On October 22, 1990, President George H. W. Bush ordered an immediate halt to military testing on Kahoʻolawe. A federal commission was established to explore how the island might be transferred back to the state. It called for clearing of all unexploded ordnance and other military debris. Congress authorized $400 million for the cleanup and granted 11 percent of the money to the state to set up an agency—the Kahoʻolawe Island Reserve Commission (KIRC)—that would eventually manage the island. The PKO had won.

ॐ

CLEARING OF THE ORDNANCE from the island in preparation for transfer to the state proceeded with all the bravado and blunder of a military operation. "All the upper administrators were military people," Higashino told me. "It's amazing that we ever win wars." Every day almost four hundred workers were shuttled out from Maui by helicopter for the day's work. Some were EOD specialists—explosive ordnance disposal. As Higashino can at a glance tell a native grass from a nonnative species, they were trained to distinguish a potentially lethal unexploded butterfly bomb from a piece of shrapnel, a live shell from a dummy one. On Maui, "Stop the Bombing" bumper stickers gave way to ones reading "It's a Blast—Kahoʻolawe Bomb Disposal."

It turned out that there were lots of live ones. They were found all over the island in an operation that in its mathematical exactitude was about as unlike the ancient ahupua'a system as possible. The entire island was gridded into one-hundred-meter squares. If necessary, crews with weed whackers and chainsaws went out first to clear away any vegetation that obscured the ground. More than once a worker heard the bright *tink!* of the weed whacker's metal flail contacting a bomb casing. None of the bombs went off. Next, a crew of eight EOD specialists walked the square in tight formation, picking up any shrapnel. If they spotted anything that looked live, they called in the bomb squad—the disposal specialists who figured out whether the explosive was safe to move or had to be detonated in situ. After that a crew came in with metal-detecting equipment able to home in on metal up to four feet below the surface. If that crew found anything, it was excavated and either removed or blown up at the site.

That was the plan, anyway. The reality was a bit grittier. The Navy promised to clear the surface of the entire island and to use the metal detectors on 25 to 30 percent. By 2000, however, the project was running out of money, and officials scaled back the goals: only 70 percent of the island would be cleared on the surface, and only about 10 percent would be cleared underground. The integrity of the clearance was always a bit in question too. Kaho'olawe's rocks and soil contain a lot of iron. Those running the metal detectors were constantly getting false positive readings and may have skipped over some real chunks of metal. Before taking the short helicopter ride out to the island from Maui, all visitors must sign a legalistic waiver with some alarming language: "Knowing that the island is dangerous and unsafe and that the pervasive presence of unexploded explosives present to me A RISK OF SERIOUS BODILY HARM OR DEATH, I nevertheless desire to go to the island of Kaho'olawe. . . . I voluntarily ASSUME THE RISK OF INJURY OR LOSS created by the presence of explosives and other hazardous conditions, which exist on the island. . . . With full knowledge of the hazards, I RELEASE AND AGREE TO INDEMNIFY AND HOLD HARMLESS the State of Hawai'i and any and all of its officers, agents, and employees, for death or injury to me or damage to or destruction of any of my property resulting from the hazardous conditions previously listed, to include transportation to, from, on or in the island and waters of Kahoʻolawe."

One day I was driving up the main island road with Kalei Tsuha, a cultural affairs specialist with the KIRC, and a couple of volunteers. Tsuha is a Native Hawaiian who has dark blue tattoos on every limb and who had participated in numerous PKO trips to the island over the years. She has two daughters. "I practically raised my daughters on Kahoʻolawe," she said. "I used to take them out of school so they could come along. It was their university. They got to camp out with doctors, with lawyers." On the drive she was telling us how the man operating the grader on the road had one day heard the sharp skid of his blade on metal. He looked down, and there was the exposed skin of a 250-pound bomb. He quit on the spot. His EOD escort gingerly used the blade to excavate the bomb, but decided that it could not be moved safely. It was fenced off while the paperwork was filled out. A 100-pound and a 500-pound bomb also turned up nearby. Drivers found themselves slaloming between the hastily erected warning fences. "It was a religious experience, believe me," Tsuha said, and then pointed out that such experiences are simply de rigueur for the KIRC staff. In such an erosive landscape munitions have a way of turning up even in places that have ostensibly been cleared. "This land will never be cleared 100 percent," she said. "You just never know. The earth shifts. Sometimes bombs come 'porpoising up,' as they call it. You just never know." The main road traverses some of the most heavily used target areas, and it had to be regraded so many times that it ended up costing a million dollars a mile, even though most of it remains unpaved.

In 2000 a couple of live bombs turned up near the Navy's camp at Honokanaiʻa, near the island's southwest tip. The workers based there were directed to leave the windows open while the bombs were detonated in place. When they came back at the end of the day there was dust everywhere, as well as a single hole in the wall of the mess hall where a piece of shrapnel had ripped in and taken out a ketchup bottle.

The Navy finally washed its hands of the place in 2003. It had removed from the island twelve million tons of scrap metal and forty thousand tires. It had finally managed to rid the place of goats by outfitting a "Judas goat" with a radio collar; sociable to a fault, other goats clustered near it, where they could be tracked and shot. The Navy left behind a good—if somewhat bomb-prone—road traversing the island, a base camp consisting of metal barracks, and a desalinization plant that could provide drinking water to workers. On November 11 it officially transferred control over access to Kahoʻolawe to the state of Hawaiʻi, which was to hold the land in trust until the formation of "the sovereign Native Hawaiian entity."

⅋

THE EXTENT OF THE EROSION PROBLEM on Kahoʻolawe has been evident for well over a century, and efforts to revegetate the island date back almost that far. In the 1880s the sheep ranchers introduced kiawe, or mesquite trees, from South America in an effort to get something to grow. They did fine, but the erosion continued. Angus MacPhee's ranchers planted all sorts of nonnative grasses and trees, and some thrived. They shot and captured an awful lot of goats too, but despite yeoman efforts, they failed to get rid of them all. Once the ranchers were evicted, the goats multiplied once again.

Not until after the PKO had started to agitate on the island's behalf did the Navy begin planting trees with the help of the Hawaiʻi Department of Land and Natural Resources. The goal was to control erosion, not reestablish native species, so the trees planted were tamarisks and casuarinas, tough, weedy species. Workers blasted holes for the seedlings by blasting explosive rounds into the hardpacked ground. They planted the trees in straight, north-south rows, perpendicular to the island's crest: when you fly over Kahoʻolawe they are among the island's most prominent features, along with the hardpan plains that run down from those lines like wide crusts of dried blood.

When you're on the ground, some of the most prominent features are the myriad plastic posts that demarcate which areas were cleared of explosives and which weren't. Yellow posts ("DANGER") mark the edges of areas—usually of rugged topography—that weren't cleared at all. The more common orange ones ("WARNING") indicate where surface clearance, but no subsurface clearance, was done. There are also large quantities of blue flagging tied around rocks, marking the many places where archaeological artifacts such as adzes, chunks of coral, and marine shells have been found, areas that, in the universal manner of agency personnel, everyone calls "arc sites." There are a lot of arc sites that mark ancient camp- or worksites, especially in the high elevations where the heavy artifacts remained as the soil was washed away from underneath them. "There are literally hundreds of them in the high areas," Rowland Reeve, an archaeologist who has done a great deal of work on Kahoʻolawe, told me. "If you look at a map of the center of the island it looks like measles, there are so many sites."

The tamarisk and casuarina trees took root pretty well in the hardpan, and they did serve as windbreaks. They gave a hint that other plants could perhaps also grow on Kahoʻolawe. As the campaign to stop the bombing accelerated, a number of people were becoming increasingly interested in re-creating Hawaiʻi's endangered native plant communities. One of them was Rene Sylva, a member of the Maui chapter of the Native Hawaiian Plant Society, who in the 1980s began to experiment with planting native shrubs and grasses on the hardpan plain south of the island's summit. That's what Paul Higashino was taking me to see as we drove down through the Martian landscape.

Sylva's group had planted twenty-three thousand seedlings of thirty-four different species of native plants. Fewer than 10 percent of the plants survived, but at our visit, almost twenty years after planting, it was clear that some were thriving. A bunchgrass called ʻakiʻaki had crawled along the ground, forming low mounds that helped stabilize the soil. Hau and ʻaʻaliʻi, shrubs that are thought to have been common at low elevations in the past, had taken hold. Several specimens of ʻaʻaliʻi, which has graceful, willowlike leaves, had grown to three feet in height and twelve feet in length; sprawled downwind of the prevailing winds, they had the classic look of the krummholz one sees at timberline.

Between the plants, though, the soil remained bare, and Higashino pointed out one of the chief problems the restoration project faced here: any seeds that happen to be produced blow away in the trade winds, bouncing along the hard ground until they reach a gully. Yet to dig holes that might trap them was both difficult and, thanks to the potential for unexploded ordnance, dangerous. Bombs or shells might lie anywhere.

Rather than dig, then, the KIRC crew was trying to build new soil on the existing surface. They were doing so by placing small bales of pili grass on the ground. At first the staff tore apart the bales and let the wind scatter the stalks and seeds; they blew away entirely. Now the KIRC crew was leaving them intact. The bales did a good job of absorbing runoff, and they also trapped seeds and windblown soil at their lower edges. It was an elegant, low-tech solution, if not an inexpensive one. On the other Hawaiian islands you could buy a bale of pili grass for $5 or $6, but by the time the bales were flown over from Moloka'i by chartered helicopter each one cost $40 or $50.

The scene in front of us looked like someone's attempt at an Earth Art project. Weather-grayed bales lay scattered all over the burnt-red landscape in straight lines, in semicircles, in squares, in small Xs. The idea was to find the patterns that resulted in the greatest retention of soil and water while using the fewest bales. "It was Derek who came up with the Xs—eight bales," Higashino said. "I told him, 'No circles.' I don't want them playing tic-tac-toe out here. But once they did write 'Mahalo'"—"thank you" in Hawaiian—"to the pilot in bales." He has idly mused about encouraging Christo to wrap swaths of Kaho'olawe's hardpan in fabric that could curtail the erosion. "Maybe he'd use Rasta colors," Higashino mused.

The bales that had been in place for a few months or more had trapped small drifts of loose soil on their windward sides. Here and there a seedling grew. It was something, a start. One of the bale squares held a dried skeleton of a plant that looked familiar to me. I looked closer and saw that it was a tumbleweed, the weedy scourge of the American West. Whatever else Kaho'olawe was, it had certainly become a cosmopolitan place. As is true of the human population of Hawai'i, immigrants from all over the world had come to be at home there, whether they were welcomed or not.

Well-meaning people have in the past offered the KIRC all kinds of help. One company on O'ahu offered to donate enough compost to cover one-quarter of the island to a depth of two inches. Higashino had to decline because there was no way to get it to where it could be used. "I get a lot of phone calls from people asking, 'Why don't you do this, or that?'" he said. "I tell them, 'Why don't you come out here to see what it's like?' Some of those people, by the end of the week, they're very quiet. They see the challenges of the bombs, the wind, the sun. We've really screwed up this earth, haven't we?"

We got back into the SUV. Higashino wanted to show me the summit of the island, where the KIRC has been concentrating much of its work. We took a different route than the one coming down, and the road grew faint. At times we seemed to lose it entirely. We lurched uphill and down. I was trying to make out which side of the orange posts around us had the sticker indicating that what was behind it had not been cleared. It wasn't always easy to make out which side was the front and which the back. I was a white-knuckled passenger. I was remembering the stories about bombs coming unearthed during rain, and it had rained hard the previous two days. Who knew what might have turned up in one of the endless expanding rills and gullies around us?

"People put the blame on the military for destroying the island," Higashino was saying. "They didn't make it better"—he pronounced it "betta," in the local fashion—"but the island was pretty well devastated before the Navy." Then he turned the wheel sharply. "Whoops, arc site," he said. Then back the other way. "Whoops, another one." Finally, the road became more defined again and we bounced on uphill. We headed for the island's summit, which is on the edge of a shallow crater.

When we drove into the crater, it was like entering a different world. Thick grasses reached higher than the roof of the car; we wound through them on a narrow road as if through an Illinois prairie. Skylarks—melodious-voiced birds from Eurasia—flitted overhead, singing beautifully. The place looked great, but it didn't look much like Higashino wanted it to. In the late 1990s the KIRC had planted thirty-five thousand seedlings of native plants inside the crater. Big enough for a half-dozen football fields, the crater has deeper soil and much greater retention of moisture than any other place on the island. The seedlings grew nicely, but the weeds did better. It didn't help when in August 1999 an EOD team detonated a five-hundred-pound bomb there. The explosion started a fire that consumed much of the crater floor. That helped the weeds more than the native species. Many of the natives were still there, but they were hidden away amidst stalks of tall guinea grass, an import from Africa.

The island has been so altered that it is in many ways ecologically unstable, subject to big swings in both flora and fauna. A few months earlier the KIRC had had to shut down its volunteer work trips for a while when the island's mouse population exploded in response to winter rains. Crawling over the shoes of workers in camp, the mice were a health hazard; in a single day the KIRC staff filled two fifty-five-gallon drums with their corpses. "This was of Biblical proportions," Mar told me. The mice weren't native. The only animals that could eat them were the island's few feral cats and *pueo*, a subspecies of the short-eared owl that may have arrived in Hawai'i after the first Polynesians did. In camp the mice could be trapped, but across the island the only thing to do was to wait for them to die out of their own accord.

At the top we ran into Derek Mar and a few other KIRC staff members. They were working, with the seventh-grade boys, at planting a natural terrace just outside the crater rim. Two of the boys' adult chaperones were using gasoline-powered augurs to drill holes in the hard soil; the boys were planting the seedlings of 'a'ali'i and *'aweoweo,* a red-stemmed plant related to amaranth. After placing each in the ground they issued it with pellets of fertilizer, a handful of mulch, and an initial shot of water. They put tiny pink flags next to all the newly planted shrubs so that they wouldn't be stepped on. Perhaps an acre had been planted. In the wind the flags looked like a field of waving pink poppies. It was possible to forget, for a moment, that there were another twenty-eight thousand acres around it that had scarcely been touched.

Standing among the pink flags, Higashino was winding up his litany of ecological damage to Kaho'olawe.

"And who knows what ecological effects the Native Hawaiians had!" he said.

"None!" thundered Mar, standing next to him, grinning a little.

"They raped and pillaged!" said Higashino. He paused, then said to me, "That's the good thing about Hawai'i, is we can all give one another a hard time. Derek's Japanese and I'm Hawaiian."

They were riding each other, but Higashino soon came back to the same theme. "We like to talk about the 'noble savage,'" he said, "but every culture has ended up raping and pillaging, and then you end up with *kapu*"—taboos or rules—"intended to fix that. When I see a native area, I think it shouldn't have nonnative species; it shouldn't have ungulates. And a native person—their perception of the forest is not how things were long ago. You go out and ask a native person to name five native trees, and they say mango, guava, and so on"—all introduced species, in other words. "It's what they know."

In Hawai'i the relationship between those who focus primarily on protecting the islands' native plants and animals and those who seek to continue or resuscitate cultural traditions has not always run smoothly. When Higashino worked on Hawai'i Island he saw how feral pigs did as much damage to the wet, high-elevation forests as goats caused in the lowland woods and shrublands of Kaho'olawe. Conservationists have tried to eradicate the pigs from preserve areas, but that has upset some residents who like to indulge in what has become a long-standing and tasty local tradition: pig hunting. It's a conflict that has not arisen to the same degree on Kaho'olawe, which has no residents of its own and which has been so degraded that virtually any restorative activity—even planting those parallel rows of nonnative tamarisks—has clear benefits.

As the island gets healthier, though, those working there may have the luxury of arguing about precisely what they are aiming for. They may have the luxury, in other words, of disagreeing about just what the role of people should be. In a place like Bermuda, where I began my travels through the landscape of restoration, there isn't any argument about that because it is easy to point to a time before humans had any effect on the island's plants and animals. Bermuda's a very unusual case, however. Almost any other place in and around North America, and the rest of the world, has been home to people for a very long time.

In the oak savannas around Chicago or the ponderosa pine forests of the Southwest reasonable arguments have been made that it makes sense to restore those places to something akin to what they were like before the arrival of a particular *kind* of people—Euro-American settlers, to be precise. That's not because those advocating restoration see anything wrong with those particular settlers, but rather because the myriad new land use practices they brought with them so clearly disrupted long-standing ecological processes.

The historical record indicates that those places were taking care of themselves pretty well at that time, but they weren't pristine. People were living there. They were hunting, setting fires, moving plants around. They were shaping the landscape. They were just doing it in ways that were generally more in consonance with natural rhythms than what the new settlers did, with their axes and livestock and imported species of plants. To call for returning ecosystems to conditions similar to those that prevailed before Euro-American settlers arrived, however, can carry this danger: it may perpetuate the myth of the noble savage. It may underscore the old and ill-conceived notion that the landscape of the New World had some sort of Edenic quality when people of the Old World got there, that it was a pristine wilderness, which is a way, really, of ignoring the role of the people who were already living there.

I spoke about this with Dennis Martinez, a forester from northern California who runs the Society for Ecological Restoration's Indigenous Peoples' Restoration Network, at a conference we attended. He was wearing a T-shirt depicting Geronimo and his rifle-toting warriors that read "Department of Homeland Security: Fighting Terrorism Since 1492." "Ecology," he said, "often doesn't recognize history, and rarely culture. But culture and ecology are the same thing. There's a great overlap."

Nowhere is that clearer than in Hawai'i, where people arrived almost two millennia ago and caused untold changes long before anyone arrived with a camera or notebook. Mar joked that Native Hawaiians didn't cause any changes in the landscape, but that is clearly not true: there's the evidence of all those birds that went extinct after the first Polynesians got there. When Captain Cook arrived in 1778 he reported that vast grasslands stretched around the lower elevations of the Hawaiian Islands because native people had burned forests for agriculture. People, after all, are mammals too, and there isn't an animal that doesn't have some effect on its environment. On a set of small islands the effects of people simply become self-evident much more quickly than on a huge continent.

"In the islands," Higashino told me, "there is enough land area that you can have areas for hunting, for gathering medicinal plants, but there should also be strictly controlled areas with more native plants." There isn't *that* much room in the islands, though, so it is hard to imagine that those arguments won't keep cropping up. For the moment, however, the thought of arguments about particular uses of the land seems almost idle amid the red hardpan and the non-native grasses of Kaho'olawe. Those who have committed themselves to healing the island find themselves adhering to the same rule as a physician: first, do no harm. Simply to staunch the island's most critical wounds is going to take an astonishing amount of work.

Higashino was eager to show me a project that was going to make it much easier: the KIRC had recently finished a water catchment system at the summit. An expanse of metal sheets more than an acre in extent caught rainwater and distributed it to tanks that could hold more than half a million gallons, which then could be fed through irrigation hoses to the new plantings. Although many of the seedlings were planted without irrigation, those that received the supplemental water had a better shot at survival in the face of the strong sun and winds. A state grant was paying for the catchment system and the seedlings. It was a sign that people outside the ranks of botanists and ecologists were recognizing the importance of the work here.

The catchment system was impressive, but as Higashino was showing it to me I could not get out of my mind something that Kalei Tsuha had said earlier. When the plans for the water-collection system were announced, she related, she and a former supervisor, both PKO members, had told the ecologists that it wouldn't work in isolation. "If you don't have a ko'a," they'd said, "you're not going to gather squat." That's when I decided I needed to experience a trip to the island with the PKO.

ॐ

WHEN HE WAS A SENIOR in Kamehameha High School on O'ahu in 1993, Derek Kekaulike Mar's class participated in the annual school singing contest. It was the one-hundredth anniversary of the U.S. takeover of Hawai'i, and most of the other classes sang songs either lamenting the loss of sovereignty or calling for its reinstitution. Instead, Mar's class chose to sing "Mele o Kaho'olawe," a song about

Kahoʻolawe composed by Uncle Harry Mitchell, who was Kimo
Mitchell's father and who served as a mentor to many of the members of the PKO (in Hawaiʻi the appellations "Uncle" and "Aunty"
are generic titles of respect for elders).

Mar did not know, back in high school, that Kahoʻolawe would
take on special meaning for him. He went to college at the University
of Hawaiʻi at Manoa, studied business law, and then switched to
zoology and Hawaiian studies. He ended up dropping out, but not
before he'd learned to speak Hawaiian. He'd learned only a smattering of the language at home as a child, but in college speaking it had
become a badge of cool.

On trips out to the sea around Kahoʻolawe with fishermen in his
family he first came to realize the island's plight. "I noticed that the
coral reefs were just dead from all the sediment runoff, and I wanted
to do something about it," Mar said. He began volunteering with the
KIRC and with the PKO. Soon he was volunteering a lot. After dropping out of school he got a restaurant job. He'd work three weeks
straight so that he could have a solid week off and work on the island.
Finally a job opened up doing revegetation work with the KIRC.

Since then Mar has spent an extraordinary amount of time on
Kahoʻolawe. He serves as a guide—a *kua*—on almost every PKO
trip. He is trained in wilderness medicine and in the identification of
hazardous ordnance. He is a teller of tales, a comforter of those who
develop blisters, a skilled operator of small boats. Two months after
my first visit to the island I stood next to him atop the summit of
Moaʻulanui again. This time, on days off from his regular job, he
was shepherding a group of about forty young Hawaiians around
the island, and he was, as is his practice on PKO trips, going by his
traditional middle name. We looked at the erosion-control blankets
I'd watched the boys stake down two months earlier. Where the thick
rolls of fibers dipped into little gullies some loose soil had accumulated, and in one of those patches a few tiny seedlings had raised
their green heads. The bridge of clouds had lowered on some days
during the winter, granting rain to the island's summit. Perhaps those
seedlings were the future of Kahoʻolawe, emerging. Perhaps those
seventh-graders had been too. Perhaps the PKO's guests were as they
listened to Kekaulike talking about the place.

"You know, it grows exponentially," he was saying. "One plant will kick off how many seeds, and those will kick off how many more? I liken it to the restoration of our Hawaiian people—it's going to take time. And it's holistic. We have to incorporate technology. We have to incorporate foreigners and people of native blood. We have to incorporate people of foreign blood born here. We have to incorporate the nonnative plants. They can serve a purpose—as windbreaks, or conditioning the soil."

Moa'ulanui isn't the only site on Kaho'olawe that's viewed as culturally significant by Native Hawaiians. Another is a stony hill about a half-mile away. Named Moa'ulaiki, it is the island's second-highest summit. Kekaulike took us there next, in part for the view and in part to reveal to us the importance of cultural traditions in the island's restoration. When we got to its base Kekaulike stopped us and told us to take off our shoes and line up in two rows, men and women.

This practice was an echo of an annual celebration that PKO participants call Makahiki. It is based on an ancient ritual that took place throughout the Hawaiian Islands when the Pleiades appeared on the horizon at sunset in autumn and again when they set there in late winter; participants walked each island, asking for rain and for a blessing of the wintertime planting season. Abandoned early in the nineteenth century, the ritual was revived on Kaho'olawe by the PKO in 1982. One of the places that PKO walkers visit is Moa'ulaiki. When they ascend the hill, they walk barefoot, Kekaulike told us now, so as to be more in touch with the place and to show humility. They move up in silence, although it is a good idea to think on a particular chant during the walk. So we walked uphill, we novitiates, shuffling unshod on feet accustomed to shoes, brushing past grasses and clinging vines and picking our way up over bare slopes covered with tiny shards of gravel.

Every place has its own etiquette of walking, I thought. Among the waving tallgrasses of Illinois it is a constant challenge to keep one's eyes open when the ripening seed heads are at eye level; on Utah's Escalante River there is a very particular way of trying to float one's feet along, swiftly, to avoid being sucked down into the quicksand. Hiking barefoot up Moa'ulaiki requires a slow and almost mincing motion, brushing the balls of the feet against the ground while lifting them to dislodge any tiny stones. Spend any time trying to get to know a place well enough so that you can work there with a degree of wisdom, and you soon learn that each place has its own particularities, its unique calls upon the human body and spirit.

Its own particularities: I am not a Native Hawaiian; I speak no Hawaiian; the traditions of the islands sound compelling to me but do not resonate in my own experience. I cannot speak to what it is like to participate in a Makahiki celebration and to see my own religious beliefs so closely tied to the healing of this particular, red-earthed, much-abused place. That is for the initiated, for those with much deeper ties there than I have. I believed Kekaulike, though, when he said that it was vitally important to him. And what I did understand, viscerally, on the walk up Moaʻulaiki was how people come to love places enough to devote days, years, even entire lifetimes to their healing. One of the meanings of the word *restore* is to put back something no longer present, and what the members of the Protect Kahoʻolawe ʻOhana are putting back are not only plants, but also the sort of stories that once enlivened the landscape. Some are old stories; some are new. Some are sacred; some can be downloaded from the Internet. They're stories of human tenancy on a particular piece of land, and they lend both the land and its people a signal depth of meaning.

I was unable to muster a chant in Hawaiian as we walked, but instead one in my own language came to me, one with a useful cadence for our slow pace: *I voluntarily ASSUME THE RISK OF INJURY OR LOSS. I voluntarily ASSUME THE RISK OF INJURY OR LOSS.* The boilerplate waiver we had all had to sign to get onto Kahoʻolawe had it right: that's exactly what happens in coming to love a place. *ASSUME THE RISK*. It might be legalistic, perhaps, but it contained a germ of poetry. Kahoʻolawe, like Nonsuch Island, like Glen Canyon, had been a place that many people had been willing to walk away from, a place considered so degraded that it was given up as lost. In each case, though, a few people had not been willing to write it off. Through a lot of hard work they'd succeeded in bringing the place back into the public consciousness. A place that had been condemned no longer was.

On Kahoʻolawe the physical successes in revegetation and erosion control are real, but modest, so far: the island still bleeds sediment into the ocean with every rain, pops old bombs out as new erosion takes place, and is home to far more nonnative than native plants. The cultural success story has arguably been far greater, as thousands of people have come to see the island not as a distant place with no relevance to their lives, but rather as a place to *care about*, a place in the heart. And opening up the heart is all about assuming the risk of injury or loss. Opening up to Kahoʻolawe, or any place, is a bit like falling in love; it is a joyous gamble, a dropping of psychic armor, that carries with it the possibility of pain as well as exaltation. There is the opportunity to see how much may be accomplished; there is also the opportunity at every moment to experience anew just how much harm has been done.

"I nevertheless desire to go to the island of Kahoʻolawe," the waiver read. I recalled something Steve Packard had told me at his house in the suburbs of Chicago: "I said to myself, 'I could do this. I could save these prairies. But if I do save them I'm getting myself into something for the long term. I'll get committed to it. . . . I won't want to let the thing fail. I'll end up with ties to a lot of people. It will tie me. Am I going to do that or not?'"

Yes, he decided, he was. He was willing to fall in love. He was willing to bind himself. Like those working for the healing of Kahoʻolawe, he desired to go, and that had made all the difference. He assumed the risk. So had everyone in this crowd of people proceeding in slow formation up the stony side of Moaʻulaiki.

We shuffled up the hill, the red stains on our bare feet only the most obvious manifestation of the way the landscape was working on us. It was affecting us through myriad other expressions of itself too: in the brightness of the sun on our necks, in the buzzing of insects, in the smell of sun-warmed bunchgrasses, in the dry scrunch of gravel underfoot. As we came on up over the summit's rocky crest we could see the islands spread before us: Lanaʻi off to the west, Molokaʻi straight ahead half-tucked behind the mountains of west Maui, the great pyramid of Haleakala off to the right, the high snowy peaks of Hawaiʻi behind us. On a clearer day one of Oʻahu's peaks would have been visible too. A long gray line of clouds stretched

over us from Haleakala, but it was high up and carried no promise
of rain today.

Around the scattered islands the waters glittered, and a thought
came to me: each of these islands is a kipuka, a small oasis of life
adrift in an enormous sea. None of the flecks of land in our view—
not Kahoʻolawe, not Maui, not even the Hawaiian islands all put
together—is big enough to allow for walking away from problems.
No wonder it was the residents of an island chain, rather than of a
huge continent, who had made the reclaiming of a military target
zone such an issue back in the 1970s; from Moaʻulaiki it is terribly
clear that there isn't any new land to make up for despoiling the old.
We stood or crouched in silence for a few minutes, and then
Kekaulike spoke. This place was where the ancient Hawaiians came
to learn, he said. From up here the islands were clearly visible, as
were the lines the currents and winds made in the channels around
Kahoʻolawe, and the stars. The relationship between them all could
be seen from here. It was a place, then, that ancient Hawaiians visited to learn about humanity's role in the cosmos.

Just below the hill's summit was a great pitted slab of rock, split
into two halves, perched atop another boulder. It was eight feet long
and covered with crusts and tufts of gray-green and mustard yellow
lichens. Next to it we could see a tripod made up of light-colored
wooden poles that supported a wooden shelf, and on and around
that were yet more offerings: corals and wrapped packets of leaves.
Kekaulike took a few steps over to the great perched slab and
pounded his fists on it. It gave off a hollow, ringing sound. It was, he
said, a bell rock that had been used to call people together. It had
been carried to its location at some time in antiquity. It was, he said,
"our Stonehenge," a place that linked the present with the deep past,
and with the future.

Just past the bell rock the ground dropped away to a hardpan
slope that had been a primary target area for the military jets. The
giant circles of white-painted tires and the bombed-out vehicles had
been cleaned up, but it would take a long time to revegetate the area.
Perhaps seeds would drift down there as the newly planted shrubs
and trees and grasses near the summit matured. Off to the other side,
at the summit, we could barely see the white edge of the new water

catchment. All Kahoʻolawe's history was visible from here. It was seamless, Kekaulike suggested. The island's history had not ended when Native Hawaiians left the island or when the bombing began. It was ongoing, and if the island today seemed rife with problems of epic scale, it was also important to realize that *today* represents a pretty limited perspective.

"Hawaiians always think a few generations ahead," he said. "We're trying to think a hundred, two hundred years down the road. I envision a dryland forest here, like it used to be. I think anyone who works in restoration comes to a crossroads where you think, 'Am I making a difference?' You have to think that the impact is forty, fifty, or more years down the road, when these trees will be full grown and reproducing. The missing element here is time. I'm not going to see a dryland forest on Kahoʻolawe, but maybe my grandchildren will."

I was a little saddened that none of us had brought offerings to present up here, but Kekaulike said we had simply by being here. "On Kahoʻolawe the best *hoʻokupu*"—offerings—"are brought up here, because this site is so sacred," he told us. "But your presence up here, if you're mindful about it, is a sort of offering too."

That, I thought, was exactly what Bill Jordan had been talking about when I'd interviewed him. Years ago he'd seen a hint of protocol, of self-consciously proper behavior, in the way tools were organized on a restoration workday in the Chicago suburbs. It was a way of placing a degree of human order on the physical world, and on human work in that world, so as to provide meaning. What was going on here was protocol taken to a much higher degree, a tight integration of the physical and the spiritual that may have been common once but has been lost in much of the modern world.

The verb *to act*, Jordan had said, has multiple meanings. We act by doing something—planting shrubs, stemming erosion—and we act by engaging in performance. On Kahoʻolawe both the simple acts of ecological restoration and the complex acts of ritual have the same degree of self-consciousness. You have to be self-aware to do both, and you have to be serious about it. On Kahoʻolawe the ritual is itself an act of restoration. Creating order in the physical world and creating meaning in the human sphere have been seamlessly wedded.

The glue that holds them together is hope for the future, the conviction that how we act matters deeply.

On the long walk back down to our camp someone asked Kekaulike about differences between the PKO and the KIRC. There have, predictably, been some strains between the two organizations; there have been changes within the PKO as it has gone from a firebrand organization mobilizing *against* something to an organization working *for* something. "KIRC is my *employer*," Kekaulike said. "The island is what I *work for*. It doesn't matter whether KIRC or PKO plants a tree. The tree doesn't care. But it is one more plant growing and one bit less erosion to happen."

⚘

ON THE ISLAND'S EASTERN SHORE is a deep indentation where floating ocean debris washes up. It must be a quirk of the ocean currents. The place is called Kanapou Bay, a name that refers to the many sharks found there. Anything that floats around the Pacific, it seems, washes up on Kanapou's rocky shore: glass net floats from Japanese fishing vessels; sneakers from Taiwan; bottles once filled with Australian beer and now, still capped, filled with murky mysterious liquid; the feathers of boobies and shearwaters; eroding chunks of Styrofoam. There are, Kalei Tsuha said, gobs of "left-sided slippers" —flip-flops—"but no right-sides. We don't know why." Even to the initiated, the island gives up its secrets only slowly.

It is the shore's quality of embracing everything that comes to it, some have suggested, that gave rise to the name *Kahoʻolawe*, which means "to take, to embrace." For many centuries the island has taken all of what has come to it: first the scatterings of life in the form of the few animals and plants that were able to cross thousands of miles of ocean to get there; one race and then another of adventurous seafarers; goats and cattle and all manner of new plants from distant continents; the explosive legacy of the most war-torn century in history. Hospitable to a fault, Kahoʻolawe has taken it all in. And these days it takes in the caring of people who live on Maui, on Oʻahu, on Kauaʻi, even on the mainland, who are, whatever the outcome of their work, compiling a new chapter in history.

What they're writing—with their planting, with their bare feet and hands, with their offerings—amounts to the renunciation of the old idea of paradise that we first encountered in Bermuda. Paradise, after all, is a place that lacks nothing. It is a place that gives everything and does no taking. It is the land of milk and honey. Once attained, it requires no ho'okupu—no offerings, no sacrifice. When the English got to Bermuda they thought they'd found it—a place where every man could live as his own king. Their false discovery was repeated, over and over, on the long course of American settlement from the Atlantic coast to the Pacific strand and beyond. Everywhere the land's riches were there for the taking. Everywhere they seemed for a while endless. Everywhere they did come to an end. Many people, disappointed, moved on to the next best place, the next lode, the big rock candy mountain, the place where the water really was supposed to taste like wine. It had to be out there somewhere. Maybe it was just over the state line, or on those distant islands way over the horizon. Some people, restless, are still looking for it.

Others, though, remained where they were, realizing that the real world does demand ho'okupu. It requires giving back. It exacts gifts of caring and of husbandry. In some cases those gifts might consist of no more than the sensitivity to leave a place alone; in others it might mean intensive projects of stewardship that will take untold future generations to carry on. Sometimes the gift is the planting of an 'a'ali'i, or of a chestnut tree; sometimes it is planting the seeds of culture or passing on a good story.

After my visit to Bermuda I continued to correspond with David Wingate about his work in conserving the cahow and restoring Nonsuch Island. At one point, he was upset about the way government officials had, in his view, pushed him to the side after his retirement. "Long-term restoration ecology projects are usually the dream child of one or a very few people who may have trouble finding successors with the same dream or determination," Wingate wrote. "I frequently hear of cases of succession problems jeopardizing long-term restoration projects in other areas of the world. I guess in the final analysis we are all just fallible human beings with a finite time in this world."

That's the rub, isn't it: however dedicated any one person or group of people may be, the myriad projects of ecological restoration

will succeed only if others take up the work and pass it on, generation after generation. It's a tall order in an uncertain world, but it
struck me that the members of the Protect Kahoʻolawe ʻOhana have
got as good a shot at filling it as anyone. They have made the work
of restoration part of their culture, and culture is the connective tissue that binds generations to one another. Kekaulike's experience on
Kahoʻolawe ties him both to his ancestors who engaged in similar
practices of cultural belief hundreds of years ago and to the generations to come who may sit in the shade of trees he planted and may,
also barefoot, themselves shuffle slowly to the summit of Moaʻulaiki.
It is seemingly ironic, but, in fact, entirely fitting, that it happened on
this particular island. It was only because the military took it over for
so long that Kahoʻolawe avoided another fate, that of becoming a
private resort for the wealthy. It was only because of the worst that
had been thrown at it that Kahoʻolawe still retained the ability to
serve as a sort of Hawaiian homeland amid a flood of change. It was
only because it had been so thoroughly written off that it presented
such a rich opportunity to restore not only a working community of
plants, as people were trying to do in so many places, but also a full
spectrum of human husbandries.

Hope, it seems, lifts its delicate green head in the most surprising
of places.

The ultimate kernel at the core of the ecological restoration movement is that every place matters, whether it be some tiny urban park
or an entire national forest, and that none is entirely beyond some
repair. Given enough dedication over enough time, even a grossly
despoiled island can be healed. No, not to what it was before, but at
least to some state of greater health that works better both for place
and people. A twenty-first-century Appalachian forest stocked with
hybrid chestnut trees will never be quite what the first American settlers saw when they came through the Cumberland Gap, but it will
arguably serve wildlife, and local people, better than a forest without
chestnuts. Kahoʻolawe will do Hawaiʻi's people more good when it is
no longer bleeding millions of tons of red sediment into the ocean.
Restoration *is* hope, plain and simple: to restore a place, or a species,
or a cultural tradition is to stake a claim that the future can be better
than the past and that people can do good in the world. In an era of

uncertainties and doubts, that simple equation is reason enough for people all over the world to engage in it.

Hawaiian historian Davianna McGregor has written about the idea of "cultural kipuka," which she defines as seedbeds from which traditional Hawaiian culture in the form of language and religious and cultural practices has flowered outward in recent decades. Kahoʻolawe, she told me, cannot be a cultural kipuka in that sense because all its direct ties to the Native Hawaiian past were ruptured by the military. But the island and the healing going into it can be a new sort of kipuka in the same way that all restoration sites can be. They are all places where links to an ecologically healthy past were broken, but where people informed by that past are trying to create healthy new traditions of stewardship that can, in time, spread outward to other islands and preserves and entire landscapes. At Nonsuch Island, the Chicago forest preserves, the forests of the Southwest, and many more places around the world people are working hard at renegotiating relationships with nature through this new sort of caring. The terrain is uncertain, the outcome of the negotiations unclear. Human relations with these places, like any that run rich and deep, are going to remain tangled. And productive. The result? Certainly not paradise, but perhaps health both physical and spiritual, for place and for people. Nature's restoration is what people have always been looking for in an imperfect world: a renewed chance to set things right.

Acknowledgments

و

Like the projects I've profiled here, the making of this book would not have been possible without the gracious help, offered freely, of a great many people. They took time from planting or weeding or writing papers or working with wood to show me their work; they aided me in my travels; they pointed me in the direction of other interesting projects; they engaged in spirited discussion about the science and art of restoration. I am indebted to them all, and hope their work prospers. Any errors that remain despite their help are my own.

I have immersed myself in the world of restoration for long enough that I am certain to be remiss in thanking someone, but want to reserve particular thanks for the following. For general discussions about restoration and its place in society, Eric Higgs, Keith Bowers, Andrew Light, Gary Paul Nabhan, Max Oelschlaeger, and especially William R. Jordan III, whose holistic perspective on restoration has been an inspiration to many. For making possible my research on the cahow and Nonsuch Island, David Wingate, Jeremy Madeiros, and Nancy Simmons. For teaching me about chestnuts, Fred Hebard, Lucille Griffin, Paul Sisco, Sam Fisher, Robert Zahner, Rose Houk, Dave Lazor, Hugh Irwin, Philip Rutter, Al Ellingboe, Charlotte Ross, Greg Eckert, and Bill Lord. In the Chicago Wilderness, Steve Packard, Linda Masters, Robert Betz, Debra Shore, Andrea Ross, Erica Friederici, Karen Rodriguez, John and Jane Balaban, Don Parker, Susanne Masi, Paul Gobster, Mark Leach, Laurel Ross, Christiane Rey, Rick Simkin, John McMartin, Mary Lou Quinn, Ray Murphy, and Paul Labus.

In writing about the ponderosa pine forests of the Southwest I have benefited greatly from my relationship with Northern Arizona University's Ecological Restoration Institute and from discussions with numerous colleagues, especially Wally Covington, Diane Vosick, H. B. "Doc" Smith, Tom Swetnam, Don Falk, Tom Sisk, Taylor McKinnon, Todd Schulke, Shaula Hedwall, Stephen Pyne, Terry Daniel, Gordon West, Rob Davis, Dennis Martinez, Dominick DellaSala, Sharon Galbreath, Jerry Engel, Steve Gatewood, Dennis Becker, Steve Buckley, Barbara Dean, James Aronson, Hal Clifford, Michelle Nijhuis, and Carol Haralson. In and around Glen Canyon I want to thank Joan Nevills-Staveley, Katie Lee, Andy Peterson, Travis Corkrum, Chris Peterson, Ericka Wells, Jason Shumaker, Richard Ingebretsen, Annette McGivney, Chuck LaRue, Pam Hyde, Andre Potochnik, Nick Melcher, Bill Wolverton, John Spence, Bill Vernieu, Bob Hart, and W. Richard Walker. My travel to and research about Kahoʻolawe would not have been possible without the generous assistance of Derek Kekaulike Mar, Paul Higashino, Kalei Tsuha, Lyman Abbott, Alani Apio, Davianna Pomaikaʻi McGregor, Kim Kuʻulei Birnie, Kathryn Wilder, Pualani Kanahele, and Rowland Reeve.

Back home, friends and colleagues commented on draft chapters and helped improve the manuscript, especially Jim Malusa, Irene Moore, Patrick Pynes, Tony Norris, Darcy Falk, Don Lago, Jack Doggett, and Susan Lamb. At Island Press, Jonathan Cobb and Emily Davis asked the right questions, honed phrases, and sharpened the book's focus. Finally, the book never would have been written without the support of my wife, Michele James, who traveled with me when she could, supported me from home, and gave unstintingly of her time and attention in helping me to clarify what I wanted to say. As they put it on Kahoʻolawe, I say mahalo to all.

Notes

᠅

Prologue

2 *Half a century earlier*: The story of the cahow's rediscovery is told in R. C. Murphy and L. S. Mowbray, "New Light on the Cahow, *Pterodroma cahow*," *Auk* 68 (1951): 266–280, and in David Wingate, "The Restoration of an Island Ecology," *Whole Earth Review* 60 (Fall 1988): 42–57.

2 *The Gulf Stream*: David B. Wingate, Todd Hass, Edward S. Brinkley, and J. Brian Patteson, "Identification of Bermuda Petrel," *Birding* 30, no. 1 (February 1998): 18–36.

2 *Petrels feed far out at sea*: David Allen Sibley, *The Sibley Guide to Bird Life & Behavior* (New York: Alfred A. Knopf, 2001), pp. 136–145.

4 *Birds arrived too*: Andrew Dobson, *A Birdwatching Guide to Bermuda* (Chelmsford, Essex, U.K.: Arlequin Press, 2002).

5 *Juan de Bermúdez*: Hudson Strode, *The Story of Bermuda* (New York: Random House, 1932), pp. 23–24.

5 *The* Sea Venture: Strode, pp. 27–29.

5 *"The fairies of the rocks"*: Strode, p. 29.

5 *John Smith*: Sir J. H. Lefroy, *The History of the Bermudaes or Summer Islands* (New York: Burt Franklin, 1964), pp. 4, 41, 13.

5 *Not majestic, but useful*: Wingate, "Island Ecology," p. 43.

6 *The island's first settlers*: Strode, *Story*, pp. 30–31.

6 *Rats arrived from England*: Wingate, "Island Ecology," p. 44.

6 *"Bottomelesse mawes"*: Lefroy, *History*, pp. 41–42.

6 *Rats, cats, and hogs*: Wingate, "Island Ecology," p. 44.

7 *"All kinds of murtheringes"*: Lefroy, *History*, p. 87.

7 *Ornithologists speculated*: Murphy and Mowbray, "New Light," p. 266.

7 *"Really only skin deep"*: Sir J. H. Lefroy, quoted in Strode, *Story*, p. 169.

7 *The source material for Shakespeare's* The Tempest: Strode, *Story*, pp. 173–177.

8 *"Jerusalem"*: Lefroy, *History*, p. 1.

8 *"Eternal spring"*: Andrew Marvell, *The Selected Poetry of Marvell* (New York: Signet, 1967), pp. 59–60.

9 *Not quite extinct*: Murphy and Mowbray, "New Light," pp. 267–268.

9 *"By Gad"*: David R. Zimmerman, *To Save a Bird in Peril* (New York: Coward, McCann & Geoghegan, 1974), p. 60.

10 *An unexpected problem arose*: David B. Wingate, "Excluding Competitors from Bermuda Petrel Nesting Burrows," pp. 93–102 in *Endangered Birds: Management Techniques for Preserving Threatened Species*, ed. Stanley A. Temple (Madison: University of Wisconsin Press, 1978).

10 *Nowhere nest farther north than on Bermuda*: Dobson, *Birdwatching Guide*, p. 132.

15 *Global contamination by DDT*: C. F. Wurster Jr. and D. B. Wingate, "DDT Residues and Declining Reproduction in the Bermuda Petrel," *Science* 159 (1968): 979–981.

16 *Yellow-fever quarantine site*: Bermuda Zoological Society, *Guide to Nonsuch Island "Living Museum" Nature Reserve* (April 2001), p. 16.

16 *William Beebe*: Bermuda Zoological Society, p. 16.

16 *Reform school for boys*: Bermuda Zoological Society, p. 16.

16 *Died of a blight*: Wingate, "Island Ecology," pp. 46–47; Tom Vesey, "When Disaster Struck," *The Bermudian* 73, no. 9 (September 2002): 10–21, 46.

17 *A wildlife sanctuary*: David B. Wingate, "The Restoration of Nonsuch Island as a Living Museum of Bermuda's Pre-colonial Terrestrial Biome," *ICBP Technical Publication* 3 (1985): 225–248.

18 *DDT poisoning*: Wurster and Wingate, "DDT Residues."

18 *"A nice garden for the tourists"*: quoted in J. A. Pollard, "Paradise Regained: Bringing an Island Back to Life," *Oceans* 18, no. 4 (July 1985): 42–49.

19-20 *A crab-eating heron had once lived there*: David B. Wingate, "Successful Reintroduction of the Yellow-crowned Night-heron as a Nesting Resident on Bermuda," *Colonial Waterbirds* 5 (1982): 104–115.

22 *Starlings did not nest in Bermuda until the 1950s*: Dobson, *Birdwatching Guide*, p. 140.

23 *Eggs of green sea turtles*: Wingate, "Island Ecology," p. 53.

23 *White-eyed vireo*: Wingate, "Island Ecology," p. 53.

24 *An old eighty-foot hulk*: Bermuda Zoological Society, *Guide to Nonsuch Island*, p. 16.

25 *A severe problem in the Everglades*: Mark Cheater, "A Florida Island Battles Green Invaders," *National Wildlife* 40, no. 3 (April/May 2002): 12–13.

Introduction

38   *A recent United Nations assessment*: Millennium Ecosystem Assessment, *Ecosystems and Human Well-being: Synthesis* (Washington, D.C.: Island Press, 2005), pp. 1-24.

38   *A thousand species of plants and animals*: U.S. Fish and Wildlife Service, Summary of Listed Species; http://ecos.fws.gov/tess_public/ TESS-Boxscore.

38   *Many more ought to be listed*: www.fws.gov/endangered/candidates/ index.htm.

38   *One-sixth of North America's area*: Norman Myers and Jennifer Kent, eds., *The New Atlas of Planet Management* (Berkeley: University of California Press, 2005), p. 38.

39   *Increasingly severe wildfires*: W. Wallace Covington and Margaret M. Moore, "Southwestern Ponderosa Pine Forest Structure: Changes Since Euro-American Settlement," *Journal of Forestry* 92, no. 1 (January 1994): 39–47; Craig D. Allen, Melissa Savage, Donald A. Falk, Kieran F. Suckling, Thomas W. Swetnam, Todd Schulke, Peter B. Stacey, Penelope Morgan, Martos Hoffman, and Jon T. Klingel, "Ecological Restoration of Southwestern Ponderosa Pine Ecosystems: A Broad Perspective," *Ecological Applications* 12, no. 5 (2002): 1418–1433.

39   *Soil erosion is rampant*: Myers and Kent, *New Atlas*, pp. 38–39.

39   *Invasive plants are simplifying*: Yvonne Baskin, *A Plague of Rats and Rubbervines: The Growing Threat of Species Invasions* (Washington, D.C.: Island Press, 2002).

39   *Sprawl and road building*: Jane Holtz Kay, *Asphalt Nation: How the Automobile Took Over America, and How We Can Take It Back* (New York: Crown, 1997); Stephen C. Trombulak and Christopher A. Frissell, "Review of Ecological Effects of Roads on Terrestrial and Aquatic Communities," *Conservation Biology* 14, no. 1 (February 2000): 18–30.

39   *Hurricane Katrina's devastation of the Gulf Coast*: Coalition to Restore Coastal Louisiana, *No Time to Lose: Facing the Future of Louisiana and the Crisis of Coastal Land Loss* (Baton Rouge, La.: 2000).

39   *A $1.5 billion project to restore its watershed*: Gretchen C. Daily and Katherine Ellison, *The New Economy of Nature: The Quest to Make Conservation Profitable* (Washington, D.C.: Island Press, 2002), pp. 61–86.

39   *$19 billion to clean up Chesapeake Bay*: Karl Blankenship, "Analysis Puts Bay Cleanup Tab at $19 Billion," *Bay Journal*, December 2002.

40   *In the Florida Everglades*: Scott Weidensaul, *Return to Wild America: A Yearlong Search for the Continent's Natural Soul* (New York: North Point Press, 2005), pp. 99–101.

40   *Salt evaporation ponds*: Carolyn Marshall, "Restoration of San

Francisco Bay Salt Ponds Is Begun," *New York Times,* July 20, 2004.

40   *"The greatest business frontier":* Storm Cunningham, *The Restoration Economy: The Greatest New Growth Frontier* (San Francisco: Berrett-Koehler, 2002), p. 289.

40   *The sixth great extinction event:* Richard Leakey and Roger Lewin, *The Sixth Extinction: Patterns of Life and the Future of Humankind* (New York: Doubleday, 1995); Niles Eldredge, *Life in the Balance: Humanity and the Biodiversity Crisis* (Princeton, N.J.: Princeton University Press, 2000).

44   *A "plain member and citizen":* Aldo Leopold, *A Sand County Almanac, and Sketches Here and There* (New York: Oxford University Press, 1949), p. 204.

Chapter 1. Smoldering at the Roots

51   *"The whole world was going to die":* Donald E. Davis, *Where There Are Mountains: An Environmental History of the Southern Appalachians* (Athens: University of Georgia Press, 2000), p. 196.

51   *"The worst thing that ever happened"* : Davis, p. 194.

51   *A far larger area than that:* E. L. Little, Jr., *Atlas of United States Trees, Volume 4: Minor Eastern Hardwoods* (Washington, D.C.: U.S. Department of Agriculture Miscellaneous Publication 1342, 1977), map 27.

51   *"Where there are mountains":* Davis, *Where There Are Mountains,* p. 11.

51   *Two hundred million acres:* Davis, p. 192.

51   *A quarter of all the trees:* Robert L. Youngs, "'A Right Smart Little Jolt': Loss of the Chestnut and a Way of Life," *Journal of Forestry* 98, no. 2 (February 2000): 17–21.

51   *Half of all the hardwoods:* Robert Zahner, *The Mountain at the End of the Trail: A History of Whiteside Mountain* (Highlands, N.C., 1994), p. 65.

52   *More than a hundred feet tall:* David M. Smith, "American Chestnut: Ill-fated Monarch of the Eastern Hardwood Forest," *Journal of Forestry* 98, no. 2 (February 2000): 12–15.

52   *Twenty-three feet and nine inches:* Thoreau's journal, June 2, 1852, quoted in David R. Foster, *Thoreau's Country: Journey Through a Transformed Landscape* (Cambridge, Mass.: Harvard University Press, 1999), p. 182.

52   *Seventeen feet in diameter:* Charlotte Hilton Green, *Trees of the South* (Chapel Hill: University of North Carolina Press, 1939), p. 120.

52   *It looked as though it had snowed:* Arthur Stupka, *Trees, Shrubs, and Woody Vines of Great Smoky Mountains National Park* (Knoxville: University of Tennessee Press, 1964), p. 45.

53 *The height of a tall man*: Thoreau's journal, October 25, 1852, quoted in Foster, *Thoreau's Country*, p. 182.

53 *Ideal telegraph pole*: Georganna Rice, Anita McCoy, Terri Webb, Cam Bond, Vivian Speed, Kim Gragg, and Dovie Green, "Memories of the American Chestnut," pp. 397-421 in *Foxfire 6*, ed. Eliot Wigginton (Garden City, N.Y.: Anchor Press, 1980), p. 401.

53 *"Perfect bowers"*: Thoreau's journal, October 25, 1852, quoted in Foster, *Thoreau's Country*, p. 182.

53 *More than four feet in diameter*: Thoreau's journal, October 25, 1852, quoted in Foster, p. 182.

53 *Could survive for six hundred years*: Ana Ronderos, "Where Giants Once Stood: The Demise of the American Chestnut and Efforts to Bring It Back," *Journal of Forestry* 98, no. 2 (February 2000): 10–11.

53 *Scottsville, Kentucky*: Gregory R. Weaver, "Chestnut Ghosts: Remnants of the Primeval American Chestnut Forest of the Southern Appalachians," *Journal of the American Chestnut Foundation* 16, no. 2 (Spring 2003): 14–18.

53 *Settlers built entire houses*: David D. Cameron, "Charlotte Ross and the Chestnut Connection," *Journal of the American Chestnut Foundation* 15, no. 2 (Spring 2002): 7–8.

53 *"Split like butter"*: Christopher Rand, *The Changing Landscape: Salisbury, Connecticut* (New York: Oxford University Press, 1963), p. 98.

53 *Grew tougher with age*: Ellsworth Barnard, "Excerpts from *In a Wild Place: A Natural History of High Ledges*," *Journal of the American Chestnut Foundation* 14, no. 1 (Summer 2000): 30–35.

53 *Made splendid fence posts*: John Alger, "The Chestnut Loggers," *Journal of the American Chestnut Foundation* 16, no. 2 (Spring 2003): 24–30.

53 *Snake fences*: P. L. Buttrick, "Commercial Uses of Chestnut," *Journal of the American Chestnut Foundation* 13, no. 1 (Spring 1999): 21–28.

53 *Sheets of chestnut wood*: Buttrick, p. 24.

53 *Items built of chestnut*: Buttrick; Chris Bolgiano, *The Appalachian Forest: A Search for Roots and Renewal* (Mechanicsburg, Penn.: Stackpole, 1998), p. 73.

54 *Boys made whistles out of chestnut bark, and hopeful young lovers inscribed their names*: Donald Swecker, "Logging and Other Memories of the American Chestnut," *Journal of the American Chestnut Foundation* 9, no. 1 (Spring 1995): 36–38.

54 *Burned hot if not long*: Barnard, "Excerpts," p. 34.

54 *The process of annealing brass*: Smith, "Ill-fated Monarch," p. 14.

54 *Boiling sap into syrup*: Barnard, "Excerpts," p. 34.

54 *Charcoal producers favored chestnut*: Smith, "Ill-fated Monarch,"

pp. 13–14; Rand, *The Changing Landscape*, p. 97.

54 *Tannin for the tanning industry*: Buttrick, "Commercial Uses," p. 24.

54 *Other uses for the tree*: Anne Frazer Rogers, "Chestnuts and Native Americans," *Journal of the American Chestnut Foundation* 16, no. 1 (Fall 2002): 24–29.

54 *Bees made excellent honey*: Rice et al., "Memories," p. 403.

54 *Boys collected them and rolled them up*: Henry Henkel Rhyne Sr., "Thoughts of Long Ago," *Journal of the American Chestnut Foundation* 16, no. 2 (Spring 2003): 20–21.

54 *A big tree could bear six thousand nuts*: F. L. Paillet and P. A. Rutter, "Replacement of Native Oak and Hickory Tree Species by the Introduced American Chestnut (*Castanea dentata*) in Southwestern Wisconsin," *Canadian Journal of Botany* 67 (1989): 3457–3469.

54 *Bears that wielded rocks*: William Lord, "William Lord's Wildlife Connection Essays, Reprised," *Journal of the American Chestnut Foundation* 8, no. 2 (Winter 1999): 12–16.

54 *High protein content*: John J. Morgan and Sara H. Schweitzer, "The Importance of the American Chestnut to the Eastern Wild Turkey," *Journal of the American Chestnut Foundation* 8, no. 2 (Winter 1999): 22–29.

55 *They made fine feed*: Davis, *Where There Are Mountains*, pp. 194–195.

55 *Ninety-two undigested chestnuts*: Davis, p. 194.

55 *Great flocks of passenger pigeons*: Joshua W. Ellsworth and Brenda C. McComb, "Potential Effects of Passenger Pigeon Flocks on the Structure and Composition of Presettlement Forests of Eastern North America," *Conservation Biology* 17, no. 6 (December 2003): 1548–1558.

55 *Stored away in salt*: Rice et al., "Memories," pp. 401–402.

55 *As by the Cherokees before them*: Rogers, "Chestnuts and Native Americans," p. 24.

55 *Three to four pounds of sugar*: Curt Davis, "The American Chestnut Tree," article on ACCF Web site: http://accf-online.org/lore.html.

55 *A single railroad station*: Davis, *Where There Are Mountains*, p. 195.

55 *"All New York goes a-nutting"*: Thoreau's journal, December 1, 1856, quoted in Foster, *Thoreau's Country*, p. 183.

56 *A man named Herman Merkel*: Joseph R. Newhouse, "Chestnut Blight," *Scientific American* 263, no. 1 (July 1990): 106–111.

56 *By 1907 it had shown up*: This and most other details of the chestnut blight's spread and early attempts at control are in George H. Hepting, "Death of the American Chestnut," *Journal of Forest History* 18, no. 3 (1974): 60–67.

56 *Millions at a time*: Scott Weidensaul, *Mountains of the Heart: A Nat-*

ural History of the Appalachians (Golden, Colo.: Fulcrum, 1994), p. 161.

57 *Shenandoah National Park*: Photos in *Journal of the American Chestnut Foundation* 13, no. 1 (Spring 1999): 32–33.

57 *An effort to defeat the park proposal*: Davis, *Where There Are Mountains*, p. 193.

57 *By 1938 they were all dying*: Stupka, *Trees, Shrubs, and Woody Vines*, p. 45.

57 *A game of leaping from log to log*: Clarence Wherry Brown, "Chestnut Skeletons and Ghosts," *Journal of the American Chestnut Foundation* 17, no. 1 (Fall 2003): 30–31.

58 *As much as thirty years after dying*: Hepting, "Death," p. 67.

58 *Liquid tannin was extracted*: Champion Fibre Company, *The Story of Chestnut Extract* (1937).

61 *Hemlock woolly adelgid*: C. Cheah, M. E. Montgomery, S. Salom, B. L. Parker, S. Costa, and M. Skinner, *Biological Control of Hemlock Woolly Adelgid*, FHTET-2004-04 (Morgantown, W.Va.: USDA Forest Service Forest Health Technology Enterprise Team, 2004).

62 *Chinese and Japanese chestnuts were much more resistant*: Hepting, "Death," p. 64.

62 *Plant breeders around 1930*: Hepting, pp. 65–66.

62 *No one knew precisely*: Gary J. Griffin, "Blight Control and Restoration of the American Chestnut," *Journal of Forestry* 98, no. 2 (February 2000): 22–27.

62 *Oaks and hickories, maples and black locusts*: Frank W. Woods and Royal E. Shanks, "Natural Replacement of Chestnut by Other Species in the Great Smoky Mountains National Park," *Ecology* 40, no. 3 (1959): 349–361.

63 *"Smoldering at the roots"*: Robert Frost, *Complete Poems of Robert Frost* (New York: Holt, Rinehart and Winston, 1963), p. 407.

64 *Wrote a letter to the professional journal*: Charles Burnham, "Blight-resistant American Chestnut: There's Hope," *Plant Disease* 65, no. 6 (1981): 459–460.

64 *Governed by only two or three genes*: C. R. Burnham, P. A. Rutter, and D. W. French, "Breeding Blight-resistant Chestnuts," *Plant Breeding Reviews* 4 (1986): 347–397.

64 *He saw remnant chestnuts bearing flowers*: Sally Conklin, "Dr. Burnham at Home," *Journal of the American Chestnut Foundation* 9, no. 1 (Spring 1995): 26–31.

72 *A chestnut stand near West Salem*: Susanne Quick, "Scientists Hope Work Will Help Rare Chestnut Grove to Endure," *Milwaukee Journal Sentinel*, March 3, 2003.

72   *Those trees readily outcompeted native black walnuts and red oaks*: Douglass F. Jacobs and Larry R. Severeid, "Dominance of Interplanted American Chestnut (*Castanea dentata*) in Southwestern Wisconsin, USA," *Forest Ecology and Management* 191 (2004): 111–120.

75   *Griffin wrote an account of her experience*: Lucille Griffin, "The Chestnut's Johnny Appleseeds," *Wall Street Journal*, October 2, 1985: 30.

76   *"Hypovirulent" strains*: Newhouse, "Chestnut Blight"; Griffin, "Blight Control."

79   *"You can't repeat the past"*: F. Scott Fitzgerald, *The Great Gatsby* (New York: Scribner's, 1925), p. 99.

79   *"The green light at the end of Daisy's dock"*: Fitzgerald, p. 159.

Chapter 2. Entering the Woods

81   *They didn't have a proper word for it*: John Madsen, *Where the Sky Began: Land of the Tallgrass Prairie* (San Francisco: Sierra Club Books, 1985), p. 7.

82   *"Entirely western, fresh and limitless"*: Walt Whitman, *Specimen Days* (Boston: David R. Godine, 1971), p. 93.

82   *"Of course it was possible"*: Ole Rölvaag, *Giants in the Earth: A Saga of the Prairie* (New York: Harper & Row, 1927), p. 425.

88   *Burnham and Jensen*: Quoted in Stephen F. Christy Jr., "To Preserve and Protect . . . ," *Chicago Wilderness* 2, no. 2 (Winter 1999): 4–8.

89–90   *At least a quarter-million square miles of tallgrass prairie; one-one-hundredth of 1 percent*: Fred Samson and Fritz Knopf, "Prairie Conservation in North America," *Bioscience* 44, no. 6 (June 1994): 418–421.

90   *"We will build up an exhibit"*: Aldo Leopold, "The Arboretum and the University," pp. 209–211 in *The River of the Mother of God and Other Essays by Aldo Leopold*, ed. Susan L. Flader and J. Baird Callicott (Madison: University of Wisconsin Press, 1991).

90   *The Civilian Conservation Corps*: Thomas J. Blewett and Grant Cottam, "History of the University of Wisconsin Arboretum Prairies," *Journal of the Wisconsin Academy of Sciences, Arts and Letters* 72 (1984): 130–144.

91   *Arboretum researchers carefully compared the success rates*: Blewett and Cottam.

91   *Fire was the single most effective means*: John T. Curtis and Max L. Partch, "Effect of Fire on the Competition Between Blue Grass and Certain Prairie Plants," *American Midland Naturalist* 39, no. 2 (1948): 437–443.

91   *The Curtis Prairie was burned*: Blewett and Cottam, "History."

91   *"The presence or absence of grassland"*: John T. Curtis, *The Vegetation of Wisconsin* (Madison: University of Wisconsin Press, 1959), p. 302.

95   *They collected two thousand seeds*: William K. Stevens, *Miracle Under the Oaks: The Revival of Nature in America* (New York: Pocket Books, 1995), p. 52.

99   *Those first seeds Packard and company had scattered came up*: Stevens, p. 52.

103   *An article for a new journal*: Steve Packard, "Just a Few Oddball Species: Restoration and the Rediscovery of the Tallgrass Savanna," *Restoration & Management Notes* 6, no. 1 (Summer 1988): 13–22.

103   *Eight-hundred-page* Plants of the Chicago Region: Floyd Swink and Gerould Wilhelm, *Plants of the Chicago Region* (Lisle, Ill.: Morton Arboretum, 1979).

104   *The approach was "dangerous"*: Packard, "Just a Few Oddball Species," p. 18.

104–105   *An obscure article*: S. B. Mead, "Catalogue of Plants Growing Spontaneously in the State of Illinois, The Principal Part near Augusta, Hancock County," *The Prairie Farmer* 6 (1846): 35–36, 60, 93, 119–122.

105   *"A Rosetta Stone for the savanna"*: Packard, "A Few Oddball Species," p. 19.

105   *A paper sharply critical of Packard*: Jon Mendelson, Stephen P. Aultz, and Judith Dolan Mendelson, "Carving Up the Woods: Savanna Restoration in Northeastern Illinois," *Restoration & Management Notes* 10, no. 2 (Winter 1992): 127–131.

106   *"Faking Nature"*: Robert Elliot, "Faking Nature," *Inquiry* 25 (1982): 81–93; reprinted pp. 71–82 in *Environmental Restoration: Ethics, Theory, and Practice*, ed. William Throop (Amherst, N.Y.: Humanity Books, 2000), p. 80.

107   *Car-deer accidents*: James H. Witham and Jon M. Jones, "Chicago Urban Deer Study," *Illinois Natural History Survey Reports* no. 265 (March 1987).

111   *Sixteen separate sites*: Debra Shore, "Controversy Erupts Over Restoration in Chicago Area," *Restoration & Management Notes* 15, no. 1 (Summer 1997): 25–31.

111   *Forty-eight sites*: Shore, p. 26.

111   *$75,000 a year*: Shore, p. 29.

111   *More than five thousand volunteers*: Laurel Ross, "The Chicago Wilderness: A Coalition for Urban Conservation," *Restoration & Management Notes* 15, no. 1 (Summer 1997): 17–24.

112   *A number of extremely rare species*: Peter Friederici, "Where the Wild Ones Are," *Chicago Wilderness* 1, no. 1 (Fall 1997): 6–9.

113 *More than two hundred thousand acres of public land*: Debra Shore, "What Is Chicago Wilderness?" *Chicago Wilderness* 1, no. 1 (Fall 1997): 3.

116 *Not alone in thinking it incongruous*: Overviews of the Chicago restoration controversy are in Shore, "Controversy Erupts"; Paul H. Gobster, "Introduction: Restoring Nature: Human Actions, Interactions, and Reactions," pp. 1–19 in *Restoring Nature: Perspectives from the Social Sciences and Humanities*, ed. Paul H. Gobster and R. Bruce Hull (Washington, D.C.: Island Press, 2000); and Alf Siewers, "Making the Quantum-Culture Leap: Reflections on the Chicago Controversy," *Restoration & Management Notes* 16, no. 1 (Summer 1998): 9–15.

116 *The front page*: Ray Coffey, "Half Million Trees May Face the Ax: DuPage Clears Forest Land to Create Prairies," *Chicago Sun-Times*, May 12, 1996.

117 *A cavalcade of similar articles*: Ray Coffey, *Chicago Sun-Times*: "Forest Dist. 'Partners' Have Shady History," May 31, 1996; "Restorationists Gnaw at Forest Picnic Area," June 4, 1996; "'Restorationists' Talk a Chic, but Vague, Game," June 7, 1996; "Forest Preserves' 'Controlled Fires' Raise Concerns," October 31, 1996; "Forest Preserve District Is Picking Our Poison," November 1, 1996.

117 *"So the North Branchers trod lightly"*: Stevens, *Miracle*, pp. 66–67.

118 *Stroger abruptly declared a moratorium*: Shore, "Controversy Erupts," p. 25.

118 *A series of public hearings*: Shore, p. 26.

118 *More than 130 speakers testified; Sheldon Altman*: Ken Keenan, "Forest Restoration Debated at Hearing," *Skokie Review*, November 14, 1996.

120 *The criticism dispiriting*: Reid M. Helford, "Constructing Nature as Constructing Science: Expertise, Activist Science, and Public Conflict in the Chicago Wilderness," pp. 119–142 in *Restoring Nature: Perspectives from the Social Sciences and Humanities*, ed. Paul H. Gobster and R. Bruce Hull (Washington, D.C.: Island Press, 2000).

120 *The board approved new guidelines*: Shore, "Controversy Erupts," p. 31.

121 *Master Steward Program*: Karen Berkowitz, "Neglect, Mismanagement Endanger Forest Preserves," *Glencoe News*, October 30, 2003: 9, 12.

121 *A $20 million deficit*: Michael Drakulich, "Reformers: District Needs Cleanup," *Glencoe News*, November 6, 2003: 10, 20, 22.

121 *A study of the county's natural areas*: Drakulich.

126 *In his 2003 book*: William R. Jordan III, *The Sunflower Forest: Ecological Restoration and the New Communion with Nature* (Berkeley: University of California Press, 2003).

127   *"Arguing for the protection of natural places"*: Jordan, p. 2.

127   *"A single act that linked engagement with total respect"*: Jordan, p. 3.

128   *"Plain members" of the land community*: Aldo Leopold, *A Sand County Almanac, and Sketches Here and There* (New York: Oxford University Press, 1949), p. 204.

129   *An essay in* Harper's: Frederick Turner, "Cultivating the American Garden," *Harper's* 271, no. 2 (August 1985): 45–52.

129   *"It always involves death"*: Frederick Turner, "Forum: She's Come for an Abortion: What Do You Say?," *Harper's* 285, no. 5 (November 1992): 53–54.

130   *People in many traditional societies perform explicit rituals*: Joseph Campbell, *The Masks of God: Primitive Mythology* (New York: Penguin, 1959).

130   *"An occasion for the creation of beauty"*: Jordan, *Sunflower Forest*, p. 173.

133   *An article he wrote on the savanna controversy*: Steve Packard, "Restoring Oak Ecosystems," *Restoration & Management Notes* 11, no. 1 (Summer 1993): 5–16.

Chapter 3. The Entrepreneurs

138   *Millions of acres*: David E. Brown, ed., *Biotic Communities: Southwestern United States and Northwestern Mexico* (Salt Lake City: University of Utah Press, 1994), pp. 43–44.

139   *"The West's prime 'Vacation-land'"*: Donald Culross Peattie, *A Natural History of Western Trees* (Boston: Houghton Mifflin, 1953), p. 79.

140   *$120 million, $175 million*: G. B. Snider, D. B. Wood, and P. J. Daugherty, "Analysis of Costs and Benefits of Restoration-based Hazardous Fuel Reduction Treatments vs. No Treatment." Northern Arizona University Forestry Department, June 13, 2003.

140   *Ever larger and more serious fires*: Stephen J. Pyne, *Tending Fire: Coping with America's Wildland Fires* (Washington, D.C.: Island Press, 2004), p. 206; William Wallace Covington, "Helping Western Forests Heal," *Nature* 408, no. 6809 (2000): 135–136.

141   *"The situation here is dramatically different"*: William R. Jordan III, *The Sunflower Forest: Ecological Restoration and the New Communion with Nature* (Berkeley: University of California Press, 2003), p. 16.

141   *A rowdy wood-frame logging town*: Platt Cline, *Mountain Town: Flagstaff's First Century* (Flagstaff, Ariz.: Northland, 1994), p. 3.

141   *A tour of the West's forests*: Linnie Marsh Wolfe, *Son of the Wilderness: The Life of John Muir* (Madison: University of Wisconsin Press, 1945), pp. 270–271; Frederick Turner, *Rediscovering America: John*

Muir in His Time and Ours (New York: Viking, 1985), pp. 300–303; Char Miller, *Gifford Pinchot: The Evolution of an American Conservationist* (Milford, Penn.: Grey Towers Press, 1992), pp. 3–33.

142   *"An almost limitless supply"*: George H. Tinker, *A Land of Sunshine: Flagstaff and Its Surroundings* (Flagstaff: Arizona Champion Print, 1887), p. 14.

142   *Cleared lumber from its abundant old-growth pines*: Donna Ashworth, *Biography of a Small Mountain* (Flagstaff, Ariz.: Small Mountain Books, 1991), pp. 83–125.

142   *Some two hundred thousand sheep*: Tinker, *Land of Sunshine*, pp. 17–18.

143   *Profits of something like 50 percent per year*: Ashworth, *Biography*, p. 149.

143   *You could stand on a cattle carcass*: Edward Land, 1934, quoted in James Rodney Hastings and Raymond M. Turner, *The Changing Mile: An Ecological Study of Vegetation Change in the Lower Mile of an Arid and Semiarid Region* (Tucson: University of Arizona Press, 1965), p. 41.

143   *The dirt lay bared*: William S. Abruzzi, "The Social and Ecological Consequences of Early Cattle Ranching in the Little Colorado River Basin," *Human Ecology* 23 (1995): 75–98; J. B. Leiburg, T. F. Rixon, and A. Dodwell, *Forest Conditions in the San Francisco Mountain Forest Reserve, Arizona*, Professional paper no. 22 (Washington, D.C.: U.S. Geological Survey, 1904).

144   *"You cannot improve upon it"*: Paul Russell Cutright, *Theodore Roosevelt: The Making of a Conservationist* (Urbana: University of Illinois Press, 1985), p. 227.

144   *"An imperative business necessity"*: Cutright, p. 210.

144   *"Men with the bark on"*: Ashworth, *Biography*, p. 268.

145   *A young forester, Gus Pearson*: Ashworth, pp. 205–214.

145   *An episodic mode of reproduction*: G. A. Pearson, "The Fort Valley Forest Experiment Station" (unpublished manuscript, Fort Valley Experiment Station Archives, Flagstaff, August 9, 1920).

145   *"Just like heaven"; "thick as the hair on a dog's back"*: Ashworth, *Biography*, pp. 211, 214.

145   *Muir dissuaded Pinchot from killing a tarantula*: Gifford Pinchot, *Breaking New Ground* (Washington, D.C.: Island Press, 1998), p. 103.

146   *The proposed damming of Hetch Hetchy Valley*: Miller, *Gifford Pinchot*, pp. 24–26.

146   *"The devil Pinchot against the angel Muir"*: Miller, p. 8.

146   *"A haze of gloom envelops the mountain land"*: John Wesley Powell, "The Non-irrigable Lands of the Arid Region," *Century Magazine* 39 (1890): 919.

148   *"Forest fires are wholly within the control of men"*: Gifford Pinchot, *The Fight for Conservation* (New York: Doubleday, Page & Co., 1910), p. 45.

148   *Loggers and sawmill crews were sent to fight forest fires*: Ashworth, *Biography*, p. 263.

148   *"It is of the first importance"*: Ashworth, p. 264.

148   *Massive fires in the northern Rockies*: Stephen J. Pyne, *Fire in America: A Cultural History of Wildland and Rural Fire* (Princeton, N.J.: Princeton University Press, 1982), pp. 242–252.

148   *The "10 a.m. policy"*: Pyne, pp. 272–287.

149   *A deer fawn, seemingly abandoned, in the nearby woods*: Flagstaff Symphony Guild, *Flagstaff 1876–1976: A Random Collection of Antique Photographs and Writings* (Flagstaff, 1976), pp. 20–21.

149   *Bambi was the first spokesanimal; a real live bear cub*: Pyne, pp. 176–177.

149   *One of the most effective advertising campaigns ever*: Pyne, p. 177.

149–150   *Most of the fires in the Southwest were small*: Pyne, *Tending Fire*, p. 206.

150   *That year Covington began working on a research project*: W. Wallace Covington, "The Evolutional and Historical Context," pp. 26–46 in *Ecological Restoration of Southwestern Ponderosa Pine Forests*, ed. Peter Friederici (Washington, D.C.: Island Press, 2003).

151   *Sackett and Dieterich took samples*: John H. Dieterich, *Chimney Spring Forest Fire History*, Research paper RM-220 (Fort Collins, Colo.: USDA Forest Service Rocky Mountain Forest and Range Experiment Station, 1980).

151   *"A glorious forest of lofty pines"*: E. F. Beale, *Wagon Road from Fort Defiance to the Colorado River* (Senate executive document 124, Thirty-fifth Congress, 1st session, 1858), p. 49.

152   *A handful had experimented*: Covington, "Context," pp. 45–46.

152   *The thick layer of downed needles around their trunks*: Stephen S. Sackett, Sally M. Haase, and Michael G. Harrington, "Lessons Learned from Fire Use for Restoring Southwestern Ponderosa Pine Ecosystems," pp. 54–61 in *Conference on Adaptive Ecosystem Restoration and Management: Restoration of Cordilleran Conifer Landscapes of North America*, ed. Wallace Covington and Pamela K. Wagner, General technical report RM-278 (Fort Collins, Colo.: USDA Forest Service, 1996).

153   *Near Flagstaff*: Peter Z. Fulé, W. Wallace Covington, and Margaret M. Moore, "Determining Reference Conditions for Ecosystem Management of Southwestern Ponderosa Pine Forests," *Ecological Applications* 7, no. 3 (1997): 895–908; Jody P. Menzel and W. Wallace Covington, "Changes from 1876 to 1994 in a Forest Ecosystem Near Walnut Canyon, North-

ern Arizona," pp. 151–172 in *Proceedings of the Third Biennial Conference of Research on the Colorado Plateau*, ed. Charles van Riper III and Elena T. Deshler (Washington, D.C.: National Park Service, 1997).

153   *At Mount Logan*: Peter Friederici, "Healing the Region of Pines: Forest Restoration in Arizona's Uinkaret Mountains," pp. 197–214 in *Ecological Restoration of Southwestern Ponderosa Pine Forests*, ed. Peter Friederici (Washington, D.C.: Island Press, 2003), p. 198.

153   *At Five Pine Canyon*: William H. Romme, Mike Preston, Dennis L. Lynch, Phil Kemp, M. Lisa Floyd, David D. Hanna, and Sam Burns, "The Ponderosa Pine Forest Partnership: Ecology, Economics, and Community Involvement in Forest Restoration," pp. 99–125 in *Ecological Restoration of Southwestern Ponderosa Pine Forests*, ed. Peter Friederici (Washington, D.C.: Island Press, 2003), p. 101.

153   *In the Jemez Mountains*: Craig D. Allen, "A Ponderosa Pine Natural Area Reveals Its Secrets," pp. 551–552 in *Status and Trends of the Nation's Biological Resources*, ed. Michael J. Mac, Paul A. Opler, Catherine E. Puckett Haecker, and Peter D. Doran (Reston, Va.: U.S. Department of Interior, U.S. Geological Survey, 1998).

153   *In poor health*: W. Wallace Covington and Margaret M. Moore, "Southwestern Ponderosa Pine Forest Structure: Changes Since Euro-American Settlement," *Journal of Forestry* 92, no. 1 (January 1994): 39–47; Craig D. Allen, Melissa Savage, Donald A. Falk, Kieran F. Suckling, Thomas W. Swetnam, Todd Schulke, Peter B. Stacey, Penelope Morgan, Martos Hoffman, and Jon T. Klingel, "Ecological Restoration of Southwestern Ponderosa Pine Ecosystems: A Broad Perspective," *Ecological Applications* 12, no. 5 (2002): 1418–1433.

154   *An alarming increase in the size and frequency of fires*: Sackett et al., "Lessons Learned," p. 55.

154   *They devoured the soil's organic matter*: Malchus B. Baker Jr., "Hydrology," pp. 161–174 in *Ecological Restoration of Southwestern Ponderosa Pine Forests*, ed. Peter Friederici (Washington, D.C.: Island Press, 2003).

154   *It was rank weeds rather than pines and native grasses*: Kerry L. Griffis, Julie A. Crawford, Michael R. Wagner, and W. H. Moir, "Understory Response to Management Treatments in Northern Arizona Ponderosa Pine Forests," *Forest Ecology and Management* 146 (2001): 239–245; J. A. Crawford, C.-H. A. Wahren, S. Kyle, and W. H. Moir, "Responses of Exotic Plant Species to Fires in *Pinus Ponderosa* Forests in Northern Arizona," *Journal of Vegetation Science* 12, no. 2 (April 2001): 261–268.

155   *The extent of the changes that had taken place*: W. Wallace Covington, Peter Z. Fulé, Margaret M. Moore, Stephen C. Hart, Thomas E.

Kolb, Joy N. Mast, Stephen S. Sackett, and Michael R. Wagner, "Restoring Ecosystem Health in Ponderosa Pine Forests of the Southwest," *Journal of Forestry* 95, no. 4 (1997): 23–29.

155 *Increased vigor and health*: Kimberly F. Wallin, Thomas E. Kolb, Kjerstin R. Skov, and Michael R. Wagner, "Seven-Year Results of Thinning and Burning Restoration Treatments on Old Ponderosa Pines at the Gus Pearson Natural Area," *Restoration Ecology* 12, no. 2 (2004): 239–247.

155 *About twice as much herbaceous vegetation*: Covington et al., "Restoring Ecosystem Health," p. 27.

157 *A federal judge issued an injunction*: Michelle Nijhuis, "Flagstaff Searches for Its Forests' Future," *High Country News* 31, no. 4 (March 1, 1999): 8–12.

160 *"Extreme logging"*: "Do You Have to Destroy a Forest to Save It?" *Arizona Daily Sun*, August 8, 2001, p. B8.

161 *One of the central players*: Peter Aleshire, "A Bare-Knuckled Trio Goes After the Forest Service," *High Country News* 30, no. 6 (March 30, 1998): 1, 8–11.

161 *More than thirty million board feet of lumber*: Philip Brick and Joey Bristol, "Gila Ground Zero: Linking Social Justice and Ecological Restoration in New Mexico's Tierra Alta," *Chronicle of Community* 4, no. 2 (Autumn 2000); http://ocs.fortlewis.edu/Stewardship/collaborative%20stewardship%20library.htm.

162 *"Reinstating a lost species"*: Pyne, *Tending Fire*, p. 103.

163 *Millions of acres in the Southwest alone*: U.S. Government Accounting Office, *Western National Forests: Catastrophic Wildfires Threaten Resources and Communities* (Testimony GAO/T-RCED-98-273, 1998).

164 *A hand in a card game*: Will C. Barnes, *Arizona Place Names* (Tucson: University of Arizona Press, 1997), p. 402.

167 *The rhetoric was heated*: Jacqueline Vaughn, "Show Me the Data: Wildfires, Healthy Forests, and Forest Service Administrative Appeals," paper prepared for presentation at the Western Political Science Association, March 27, 2003, Denver, Colo.; http://oak.ucc.nau.edu/js53/conf.htm.

167 *Within the perimeter of the Rodeo-Chediski fire*: Barb A. Strom, Pre-fire Treatment Effects and Post-fire Forest Dynamics on the Rodeo-Chediski Burn Area, Arizona. M.S. thesis, Northern Arizona University, Flagstaff, 2005.

167 *Grasslands or shrublands*: Melissa Savage and Joy N. Mast, "How Resilient Are Southwestern Ponderosa Pine Forests After Crown Fires?" *Canadian Journal of Forest Research* 35, no. 4 (April 2005): 967–977.

171 *Silver City was hurting economically*: Brick and Bristol, "Gila Ground Zero."

Chapter 4. The Voyage of the *Moon-Eyed Horse*

183 *It only rains six inches per year*: www.city-data.com/city/Page-Arizona.html.

183 *Stemmed from Arkansas*: Russell A. Martin, *A Story That Stands Like a Dam: Glen Canyon and the Struggle for the Soul of the West* (New York: Henry Holt, 1989), p. 187.

184 *Fantasized about blowing the dam to bits*: Edward Abbey, *Desert Solitaire: A Season in the Wilderness* (New York: Ballantine, 1971), p. 188; *The Monkey Wrench Gang* (New York: Avon, 1976), pp. 1–7.

184 *A famous publicity stunt*: Jared Farmer, *Glen Canyon Dammed: Inventing Lake Powell and the Canyon Country* (Tucson: University of Arizona Press, 1999), p. xiii.

184 *A cassette tape of songs*: Katie Lee, *Colorado River Songs* (Jerome, Ariz.: Katydid Books & Music, 1988).

188 *Together on a boat trip*: John McPhee, *Encounters with the Archdruid* (New York: Farrar, Straus and Giroux, 1971), pp. 197, 198, 203, 240.

189 *"Nature's altar"*: Frank L. Griffin, "Visit to a Drowning Canyon," *Audubon*, January 1966, reprinted pp. 82–90 in *The Glen Canyon Reader*, ed. Matthew Barrett Gross (Tucson: University of Arizona Press, 2003), p. 88.

189 *"The past is never dead"*: William Faulkner, *Requiem for a Nun*, pp. 471–664 in *Novels 1942–1954* (New York: Library of America, 1994), p. 535.

190 *Trickle to torrent*: Duncan T. Patten, David A. Harpman, Mary I. Voita, and Timothy J. Randleb, "A Managed Flood on the Colorado River: Background, Objectives, Design, and Implementation," *Ecological Applications* 11, no. 3 (June 2001): 635–643.

190–191 *"It is a runaway"*: Reuel Leslie Olson, *The Colorado River Compact* (Ph.D. thesis, Harvard University, Cambridge, Mass., 1926), p. 5.

191 *The Colorado River Compact*: Peter Friederici, "Stolen River: The Colorado and Its Delta Are Losing Out," *Defenders* 73, no. 2 (Spring 1998): 10–18, 31–33.

192 *"Altering the geography of a region"*: Quoted in Scott K. Miller, "Undamming Glen Canyon: Lunacy, Rationality, or Prophecy?" *Stanford Environmental Law Journal* 19, no. 1 (January 2000): 21–207, p. 139.

192 *It released a report*: U.S. Bureau of Reclamation, *The Colorado River: A Natural Menace Becomes a National Resource* (Washington, D.C.: U.S. Government Printing Office, 1946).

192 *Conservationists were outraged less*: This history is detailed in Martin, *Story*, and Miller, "Undamming."

193   *"Would not have any objection"*: Quoted in Miller, pp. 146–147.
193   *Brower found out too late*: Miller, "Undamming," p. 150; Eliot Porter, *The Place No One Knew* (San Francisco: Sierra Club, 1963).
194   *The flimsiest of vinyl rafts*: Abbey, *Desert Solitaire*, p. 175.
194   *"One of the bitterest lessons"*: Quoted in Kelly Duane, *Monumental: David Brower's Fight for Wild America* (San Francisco: Loteria Films, 2005).
194   *Moved into far northern Arizona with a vengeance, and details following*: Martin, *Story*, pp. 87ff.
194   *A four-lane highway from Phoenix to Chicago*: Harry H. Gilleland, *Jewel of the Colorado: The Last Facet Is Completed* (Denver: U.S. Department of the Interior Water and Power Service, 1981), p. 4.
195   *January 21, 1963*: Martin, *Story*, pp. 7–12.
195   *Arizona senator Barry Goldwater*: Martin, p. 260.
195   *"Should We Also Flood the Sistine Chapel"*: Miller, "Undamming," pp. 154–155.
195   *The spending of $1.3 billion*: Martin, *Story*, p. 284.
195   *National Environmental Policy Act*: Miller, "Undamming," pp. 157–158.
196   *The nests the great blue herons had occupied*: Tad Nichols, *Glen Canyon: Images of a Lost World* (Santa Fe: Museum of New Mexico Press, 1999), p. 35.
196   *Some of Brower's friends worried*: Farmer, *Glen Canyon Dammed*, p. 145.
196   *A marina had been built at Wahweap*: Farmer, p. 113.
196   *Newspapers and travel magazines*: For example, Walter Meayers Edwards, "Lake Powell: Waterway to Desert Wonders," *National Geographic* 132, no. 1 (July 1967): 44–75.
196   *Visited the nascent recreation area*: Edwards, p. 44; Farmer, p. 162.
196   *A glossy color booklet*: U.S. Bureau of Reclamation, *Lake Powell: Jewel of the Colorado* (Washington, D.C., 1965).
197   *The one senatorial vote he regretted*: Martin, *Story*, p. 325.
197   *Lake Foul or Blue Death*: Ellen Meloy, "Travels with Seldom," pp. 159–169 in *The Glen Canyon Reader*, ed. Matthew Barrett Gross (Tucson: University of Arizona Press, 2003), p. 160.
198   *"The most distinctive ichthyofauna"*: Wendell Minckley, "Native Fishes of the Grand Canyon Region: An Obituary?" pp. 124–177 in *Colorado River Ecology and Dam Management* (Washington, D.C.: National Academy Press, 1991), p. 128.
198   *Two hundred people a year were running the Grand Canyon; more than sixteen thousand*: Miller, "Undamming," p. 191.
199   *The old Colorado River delta*: Friederici, "Stolen River"; Charles

Bergman, *Red Delta: Fighting for Life at the End of the Colorado River* (Golden, Colo.: Fulcrum, 2002).

199   *63 percent fewer juvenile humpback chub*: Associated Press, "Fish Numbers Drop After Grand Canyon Flood," *New York Times*, March 7, 2005.

200   *"Playing God"*: Sandra Blakeslee, "In Bold Experiment at Canyon, a River Runs Through It," *New York Times*, November 23, 2004.

200   *A mighty blunt tool*: Grand Canyon Monitoring and Research Center, *The State of Natural and Cultural Resources in the Colorado River Ecosystem* (Flagstaff, Ariz., 2005).

200   *An Ed Abbey story*: Abbey, *Desert Solitaire*, pp. 157–172.

202   *Edward Abbey had suggested just that*: Edward Abbey, *Slickrock: Endangered Canyons of the Southwest* (New York: Sierra Club, Scribner's, 1971), p. 69.

202   *Fired as the group's executive director*: McPhee, *Archdruid*, pp. 208–220.

202   *November 16, 1996*: Miller, "Undamming," p. 122.

202   *"The silliest proposal"*: Miller, p. 123.

203–204   *Hundreds of millions of dollars*: Farmer, *Glen Canyon Dammed*, p. 185.

204   *Something like $100 million*: Theo Stein, "Drought Draining Power," *Denver Post*, July 5, 2004.

204   *Drought set in*: Data from the Bureau of Reclamation; http://137.77.133.1/uc/water/crsp/cs/gcd.html.

204   *The surface of the reservoir was down about ninety feet*: Jim Stiles, "Take It or Leave It," *Canyon Zephyr* 17, no. 1 (April/May 2005): 2–3.

205   *"Glen Canyon is the future"*: Richard Ingebretsen, "Lake Powell: The 'One Trick Pony'?" *Canyon Zephyr* 17, no. 1 (April/May 2005): 14–15.

205   *Hydrologists released a study*: Robert A. Young, "Coping with a Severe Sustained Drought on the Colorado River: Introduction and Overview," *Water Resources Bulletin* 31, no. 5 (October 1995): 779–788.

205–206   *A two- to three-decade period of drought*: Melanie Lenart, "Will the Drought Continue?" *Southwest Climate Outlook* (March 2005): 2–5.

206   *A team of climate scientists published results*: Tim Barnett, Robert Malone, William Pennell, Detlet Stammer, Bert Semtner, and Warren Washington, "The Effects of Climate Change on Water Resources in the West: Introduction and Overview," *Climatic Change* 62 (2004): 1–11.

207   *A sixty-four-day-long expedition*: Barry M. Goldwater, "An Odyssey

of the Green and Colorado Rivers: The Intimate Journal of Three Boats and Nine People on a Trip Down Two Rivers," *Arizona Highways* 17, no. 1 (January 1941): 6–13, 30–37.

207　*He gave Ernie Pyle a good dunking*: Ernie Pyle, *Home Country* (New York: William Sloane Associates, 1947), p. 401.

209　*Eighty-eight volunteers*: National Park Service, Trash Tracker statistics FY 04; www.nps.gov/glca/tracker/tthome.htm.

210　*It smells a bit like fragrant wine*: Rose Houk, "The Poop on the Pleistocene," pp. 14–15 in *Earth Notes: Exploring the Southwest's Canyon Country from the Airwaves*, ed. Peter Friederici (Grand Canyon, Ariz.: Grand Canyon Association, 2005).

210　*Mormon settlers lowered their wagons*: C. Gregory Crampton, *Ghosts of Glen Canyon: History Beneath Lake Powell* (Salt Lake City: Tower Productions, 1986), pp. 56–58.

211　*"Frank's music"*: Katie Lee, *Sandstone Seduction: Rivers and Lovers, Canyons and Friends* (Boulder, Colo.: Johnson Books, 2004), p. 114.

211　*"The color was intense"*: Lee, p. 114.

212　*A faraway droning of bagpipes*: Katie Lee, *All My Rivers Are Gone: A Journey of Discovery Through Glen Canyon* (Boulder, Colo.: Johnson Books, 1998), pp. 198–199.

212　*"It gets so weird in here sometimes"*: Lee, *All My Rivers*, p. 112.

215　*Sediment more than three miles deep*: Bergman, *Red Delta*, p. 69.

215　*Could have filled the Rose Bowl*: Miller, "Undamming," pp. 125–126.

216　*From one hundred and fifty to a thousand years*: Martin, *Story*, p. 310; Miller, "Undamming," pp. 171–172; Thomas Myers, *Sediment Hydrology on the Colorado River: The Impacts of Draining Lake Powell* (Flagstaff, Ariz.: Glen Canyon Institute, 1998).

219　*Reservoirs tend to promote the growth of tamarisk*: N. LeRoy Poff, J. David Allan, Mark B. Bain, James R. Karr, Karen L. Prestegaard, Brian D. Richter, Richard E. Sparks, and Julie C. Stromberg, "The Natural Flow Regime: A Paradigm for River Conservation and Restoration," *BioScience* 47, no. 11 (December 1997): 769–784.

223　*"It was possible to savor the* overview": Lee, *All My Rivers*, p. 70.

223　*Cass Hite*: Gary Topping, *Glen Canyon and the San Juan Country* (Moscow: University of Idaho Press, 1997), pp. 119–125.

223　*A steam-powered dredge*: Crampton, *Ghosts*, p. 80.

224　*"This cup of coffee cost me $5,000"*: Crampton, p. 82.

224　*Uranium miners*: Crampton, p. 98.

224　*Sentinel Rock*: Crampton, p. 22.

224　*Air Force bombers flew low overhead*: Elizabeth Sprang, *Good-Bye River* (Las Cruces, N.M.: Kiva Press, 1992), p. 12.

224   *Dumped their garbage*: Lee, *All My Rivers*, p. 35.

224   *They set enormous piles of driftwood alight*: Wallace Stegner, *The Sound of Mountain Water* (New York: Doubleday, 1969), pp. 108–109.

224   *Norm Nevills lit bonfires*: Nancy Nelson, *Any Place Any Time Any River: The Nevills of Mexican Hat* (Flagstaff, Ariz.: Red Lake Books, 1991), p. 36.

225   *Fresh steaks and ice cream*: Sprang, *Good-Bye River*, p. 18.

225   *"You feel mighty free and easy and comfortable"*: Mark Twain, *Adventures of Huckleberry Finn* (Garden City, N.Y.: Doubleday, 1985), p. 136.

225   *"My anxieties have vanished"*: Abbey, *Desert Solitaire*, p. 176.

225   *"When the world was young"*: Stegner, *Mountain Water*, p. 120.

226   *"Pooping in my own parlor"*: Lee, *All My Rivers*, p. 71.

226   *"Never allow reality to form a patina"*: Lee, *All My Rivers*, p. 191.

229   *The upper basin states argued*: Shaun McKinnon, "Feds Will Referee Colorado River Fight," *Arizona Republic*, April 21, 2005.

229   *On May 2 she announced her decision*: Shaun McKinnon, "Powell Portion Will Flow to Lake Mead," *Arizona Republic*, May 3, 2005.

230   *"West-running Brook"*: Robert Frost, *Complete Poems of Robert Frost* (New York: Holt, Rinehart & Winston, 1963), pp. 327–329.

231   *"The past is a sunlit country morning"*: Howard Mansfield, *The Same Ax, Twice: Restoration and Renewal in a Throwaway Age* (Hanover, N.H.: University Press of New England, 2000), p. 62.

232   *"This was only one of a hundred such places"*: Frederick Dellenbaugh, *A Canyon Voyage* (1908), excerpted pp. 34–45 in *The Glen Canyon Reader*, ed. Matthew Barrett Gross (Tucson: University of Arizona Press, 2003), p. 38.

232   *As much chance of decommissioning Glen Canyon Dam*: Laura Paskus, "Glen Canyon Dam Will Stand," *High Country News* 37, no. 19 (October 17, 2005): 6.

232–233   *Lake Powell would perhaps never fill up*: David R. Brower, "Perspective for the Colorado Canyons," *Sierra Club Bulletin* 48, no. 9 (December 1963): 66–88.

Chapter 5. Under the Bridge of Clouds

238   *Scientists knew of the po'ouli*: Paul E. Baker, "Status and Distribution of the Po'ouli in the Hanawi Natural Area Reserve Between December 1995 and June 1997," *Studies in Avian Biology* 22 (2001): 144–150.

238   *About seventy bird species*: Paul C. Banko, Reginald E. David, James D. Jacobi, and Winston E. Banko, "Conservation Status and Recovery Strategies for Endemic Hawaiian Birds," *Studies in Avian Biology* 22 (2001): 359–376.

240   *Large numbers of land snails are known to have vanished from the islands*: David A. Burney, Helen F. James, Lida Pigott Burney, Storrs L. Olson, William Kikuchi, Warren L. Wagner, Mara Burney, Deirdre McCloskey, Delores Kikuchi, Frederick V. Grady, Reginald Gage II, and Robert Nishek, "Fossil Evidence for a Diverse Biota from Kaua'i and Its Transformation Since Human Arrival," *Ecological Monographs* 71, no. 4 (November 2001): 615–641.

240   *Ten percent of the archipelago's plants*: Linda W. Cuddihy and Charles P. Stone, "Summary of Vegetation Alteration in the Hawaiian Islands," pp. 467–472 in *A Natural History of the Hawaiian Islands: Selected Readings II*, ed. E. Alison Kay (Honolulu: University of Hawaii Press, 1994), p. 468.

240   *A third of the plant species on the federal threatened and endangered species list are Hawaiian*: Nellie Sugii and Charles Lamoureux, "Tissue Culture as a Conservation Method: An Empirical View from Hawaii," pp. 189–205 in *Ex Situ Plant Conservation: Supporting Species Survival in the Wild*, ed. Edward O. Guerrant Jr., Kayri Havens, and Mike Maunder (Washington, D.C.: Island Press, 2004), p. 189.

240   *Around the time of Christ*: David E. Stannard, *Before the Horror: The Population of Hawai'i on the Eve of Western Contact* (Honolulu: University of Hawaii Social Science Research Institute, 1989), p. 32.

240   *By the early eleventh century*: William Barrera Jr., "Kaho'olawe Archaeology: An Overview," *Hawaiian Archaeology* 1, no. 1 (1984): 31–43.

240   *Changed to such a degree in the last two hundred years*: Cuddihy and Stone, "Summary of Vegetation Alteration."

241   *People burned areas*: Patrick V. Kirch, "The Impact of the Prehistoric Polynesians on the Hawaiian Ecosystem," *Pacific Science* 36, no. 1 (1982): 1–14; J. T. Tunison, C. M. D'Antonio, and R. K. Loh, "Fire and Invasive Plants in Hawai'i Volcanoes National Park," pp. 122–131 in *Proceedings of the Invasive Species Workshop: The Role of Fire in the Control and Spread of Invasive Species* (Tallahassee, Fla.: Tall Timbers Research Station Miscellaneous Publication no. 11, 2001).

241   *"Neither houses, trees, nor any cultivation"*: Kaho'olawe: Na Leo o Kanaloa (Honolulu: 'Ai Pohaku Press, 1995), p. 112.

242   *"Nearly over-run with weeds, and exhausted of their inhabitants"*: Kaho'olawe: Na Leo o Kanaloa, p. 112.

242   *At least four hundred thousand*: Stannard, *Before the Horror*.

242   *Abolished the ancient* kapu *system*: Davianna Pomaika'i McGregor, "The Cultural and Political History of Hawaiian Native People," pp. 333–381 in *Our History, Our Way: An Ethnic Studies Anthology*, ed. Gregory Yee Mark, Davianna Pomaika'i McGregor, and Linda A. Revilla (Dubuque, Iowa: Kendall/Hunt, 1996).

242   *A delegation from Boston arrived in 1820*: Lawrence H. Fuchs, *Hawaii Pono: An Ethnic and Political History* (Honolulu: Bess Press, 1961), pp. 7–17.

243   *Kahoʻolawe was turned into a penal colony*: *Kahoʻolawe: Na Leo o Kanaloa*, p. 112.

243   *Two thousand sheep*: *Kahoʻolawe: Na Leo o Kanaloa*, p. 112.

243   *Twenty thousand sheep*: *Kahoʻolawe: Na Leo o Kanaloa*, p. 112.

243   *"Entirely denuded of top soil"*: *Kahoʻolawe: Na Leo o Kanaloa*, p. 112.

243   *So many sandalwood trees were cut down*: Michael G. Buck, "Introduction: Rains Always Follow the Forest," pp. xiii–xxiii in *Wao Akua: Sacred Source of Life* (Honolulu: State of Hawaiʻi Department of Land and Natural Resources, Division of Forestry and Wildlife, 2003), p. xv.

243   *Vast swaths of forest . . . were cleared*: Buck, p. xvii.

243   *Water supplies in the lowlands had become irregular, floods more damaging*: Buck, pp. xv–xvii.

243   *A territorial forest reserve system*: Buck, pp. xvii–xix.

243   *Declared the entire island a forest reserve*: *Kahoʻolawe: Na Leo o Kanaloa*, p. 113.

244   *Leased the entire island as pasture for cattle and horses*: Inez MacPhee Ashdown, *Recollections of Kahoʻolawe* (Honolulu: Topgallant, 1979), p. 8.

244   *He was from Wyoming*: Ashdown, p. 4.

244   *His cowboys captured thirteen thousand goats*: Ashdown, p. 11.

244   *"The land smelled horribly of death"*: Ashdown, p. 11.

244   *MacPhee's efforts to revegetate the island*: Ashdown, p. 15.

244   *His family raised watermelons and papayas and eggplants*: Ashdown, p. 15.

244   *Horses and cattle raised on Kahoʻolawe brought a profit*: Ashdown, p. 60.

245–246   *Likened restoring the island's hardpan to revegetating the Costco parking lot*: Office of Hawaiian Affairs, "What's Next for Kahoʻolawe?" November 29, 2003; www.oha.org/cat_content.asp?contentid=138&catid=57.

246   *Almost two million tons of it*: Kahoʻolawe Island Reserve Commission, *Hoʻola Hou i Ke Kino o Kanaloa: Kahoʻolawe Environmental Restoration Plan* (May 1998), p. 15.

246   *Allowed the Army Air Corps*: Ashdown, *Recollections*, p. 65; *Kahoʻolawe Island: Restoring a Cultural Treasure*, p. 119.

246   *The military declared martial law*: Ashdown, p. 65; *Kahoʻolawe: Na Leo o Kanaloa*, p. 113.

246   *The MacPhee family was never compensated*: Ashdown, pp. 67–70.

247   *He thought he might import pigs*: Ashdown, p. 67.

247   *Signed an executive order*: Quoted in *Kahoʻolawe Island: Restoring a Cultural Treasure* (Final Report of the Kahoʻolawe Island Conveyance Commission to the Congress of the United States, March 31, 1993), p. 110.

247   *The Navy threw everything it had*: *Restoring a Cultural Treasure*, pp. 129–130.

248   *Operation Sailor Hat*: *Restoring a Cultural Treasure*, p. 26.

248   *Hawaiian culture seemed down and out*: Rodney Morales, "George Helm—the Voice and Soul," pp. 10–33 in *Hoʻihoʻi Hou: A Tribute to George Helm & Kimo Mitchell*, ed. Rodney Morales (Honolulu: Bamboo Ridge Press, 1984).

248   *Nine people jumped out*: Morales, p. 18.

249   *"Blow out the candles on the white man's two-hundredth birthday cake"*: Charles Maxwell Sr., "A Historical Perspective," www. kahoolawe. org/perspective.html.

249   *The Protect Kahoʻolawe Association*: Morales, "George Helm," pp. 19–23.

249   *Sixty-five Hawaiians landed on Kahoʻolawe*: Morales, p. 20.

249   *Group members soon filed several lawsuits*: *Restoring a Cultural Treasure*, p. 27.

249   *Aloha ʻaina*: Morales, "George Helm," pp. 19–20.

250   *January 30, 1977*: Morales, pp. 23–24.

250   *A rising star*: Morales, *Hoʻihoʻi Hou*.

250   *"We are motivated to pursue the action"*: Quoted in Morales, p. 69.

250   *"The island is 45 square miles"*: Quoted in Morales, p. 65.

250   *A mass invasion of the island*: Morales, p. 26.

250   *Early on the morning of March 6*: Morales, pp. 30–32; Rodney Morales, "Kimo Mitchell—a Life," pp. 76–83 in Morales.

251   *Many thought he'd been murdered*: For a fictionalized treatment of this idea, see Rodney Morales, *When the Shark Bites* (Honolulu: University of Hawaii Press, 2002).

251   *The campaign to stop the bombing*: *Restoring a Cultural Treasure*, p. 27.

251   *October 22, 1990*: *Restoring a Cultural Treasure*, p. 111.

251   *A federal commission*: *Restoring a Cultural Treasure*.

254   *"The sovereign Native Hawaiian entity"*: Hawaii Revised Statutes, chapter 6K.

254   *Kiawe, or mesquite trees*: *Restoring a Cultural Treasure*, p. 23.

255   *Sylva's group had planted*: Alan D. Ziegler, Allison Chun, John L. Perry, John R. Egan, Kanani Garcia, and Thomas W. Giambelluca, "Assessment of the Native Hawaiian Plant Society Restoration Projects

on Kaho'olawe Island, Hawai'i," *Ecological Restoration* 18, no. 1 (Spring 2000): 26–33.

258   Pueo: S. L. Olson and H. F. James, "Prodromus of the Fossil Avifauna of the Hawaiian Islands," *Smithsonian Contributions to Zoology* no. 365 (Washington, D.C.: Smithsonian Institution, 1982).

260   *When Captain Cook arrived in 1778*: Buck, "Rains Always Follow the Forest," p. xiv.

261   *"Mele o Kaho'olawe"*: Quoted in Morales, *Ho'iho'i Hou*, pp. 86–87.

263   *The ritual was revived on Kaho'olawe by the PKO in 1982*: Pualani Kanaka'ole Kanahele, "Ke Au Lono I Kaho'olawe, Ho'i (The Era of Lono at Kaho'olawe, Returned)," pp. 125–141 in *The Colors of Nature: Culture, Identity, and the Natural World*, ed. Alison H. Deming and Lauret E. Savoy (Minneapolis: Milkweed, 2002); Bert Lum, "Kaho'olawe and the Makahiki Ceremony: The Healing of an Island," *Californian Journal of Health Promotion* 1 (2003): 25–33.

268   *"To take, to embrace"*: Kaho'olawe: Na Leo o Kanaloa, p. 43.

271   *Cultural and economic influence*: Davianna Pomaika'i McGregor, "Waipi'o Valley, a Cultural Kipuka in Early 20th Century Hawai'i," *Journal of Pacific History* 30, no. 2 (1995): 194–209; *Na Kau'aina: Keepers of Hawai'i's Sacred Nature* (Honolulu: University of Hawaii Press, in press).

Index

Note: page numbers in *italics* refer to illustrations